HIGH ADVENTURE

HIGH ADVENTURE

A.H.COBBY
D.S.O. D.F.C. & Bars

KOOKABURRA TECHNICAL PUBLICATIONS PTY. LTD.
PO BOX 648, DANDENONG 3175,
MELBOURNE, VICTORIA, AUSTRALIA

FIRST PUBLISHED IN 1981

ISBN 0 85880 044 6

PLANNED AND DESIGNED IN AUSTRALIA
BY KOOKABURRA TECHNICAL PUBLICATIONS PTY. LTD.
PRINTED IN SINGAPORE BY KOON WAH PRINTING PTE. LTD.

Foreword

This rare photograph shows Air Marshal Sir George Jones as a captain and flight commander in No. 4 Squadron, Australian Flying Corps in 1918.

This is the story of Harry Cobby, the most successful and most highly decorated fighter pilot of the Australian Flying Corps in the war of 1914-18. He destroyed 24 German aircraft, drove down another five out of control, and also destroyed several balloons. It was my privilege to serve with him in No. 4 Squadron in 1918 and I learned to know his great qualities as a fighter pilot and as a leader of men. He was a few years older than most of us and this, together with his other attributes, made him our natural leader in the air and in all off-duty activities. He was a man of great good humor and ready wit, as is well shown in this story.

We had no parachutes in those days and fire or the destruction of a vital part of an aircraft meant certain death. Cobby repeatedly sought out enemy aircraft, and whether as a member of a formation or alone, fought many duels with them. For this he was greatly admired.

To fight a duel in the air required considerable courage, especially if you were flying alone, and with only your self respect to drive you on. When fighting as a member of a formation it was much easier because your conduct was almost certain to be observed by others and any sign of cowardice called forth harsh comment. In all cases the nervous tension was enormous. Several times pilots were unable to face the situation and they had to be sent back to England.

After the war, Harry Cobby became a foundation member of the Royal Australian Air Force and commanded R.A.A.F. Base Richmond, NSW, for some years. He later transferred to the Civil Aviation Department. On the outbreak of war in 1939, he returned to the R.A.A.F. and did splendid work as Director of Recruiting and later as Commander of the First Tactical Air Force at Noemfoor and Morotai Islands in the East Indies.

His service to Australia is an outstanding example of courage and devotion to duty, as shown when he saved a fellow officer from drowning after a flying boat accident at Townsville. For this he was awarded the George Cross. As a prominent member of the Return Services League, his untimely death on Armistice Day, 1955, was a sad loss.

George Jones

Air Marshal Sir George Jones, K.B.E., C.B., D.F.C. (Retd.)

Acknowledgements

This book has an unusual history. Published originally in 1942 when the author was the R.A.A.F.'s Director of Recruiting, it was virtually a straight autobiography with only a few photographs. It nevertheless became a recognised classic. Many years later Norman Clifford, an experienced commercial artist and W.W. 1 aircraft historian, who had held the original manuscript in his possession since 1947, approached David Cobby, the author's grandson, and suggested a reprinted edition. This has finally come to fruition in a new semi-pictorial form as part of the celebrations of the R.A.A.F.'s 60 Years Diamond Jubilee in 1981.

The Cobby family's photographic albums proved to be a vital and virtually untapped source for many of the photographs. In addition, many other previously unpublished shots were located, even as late as 1981, through the relatives of those who had served with Nos. 2 and 4 Squadrons, Australian Flying Corps. The R.A.A.F. Museum Collection at Point Cook was also a source of priceless material, so that it is now possible to present a fascinating historical coverage of the life and times of Australia's pioneer air aces such as Arthur Henry Cobby.

The Publishers would especially like to thank Richard Walsh of Angus & Robertson Publishers, who kindly gave permission for this edition to be produced. Finally, Norman Clifford, the compiler and researcher of this production — as well as its illustrator — would especially like to thank Mrs. A. Murray Jones for the photographs concerning her husband and also for the *Kaiser-i-Hind* series in the last chapter; Mrs. Joan Bourne, daughter of 'Tab' Pflaum, for her trust in sending from South Australia, two albums of her late father; Mr. R. deC. Tunbridge; Sir Ronald East; Air Marshal Sir George Jones, K.B.E., C.B., D.F.C. (Retd.), for his kindness in writing the foreword; Wing Commander Eric Roberts (Retd.); Mr. Jack Docherty, Editor of R.A.A.F. News; the Photographic Section of the Australian War Memorial; the staff of the R.A.A.F. Museum at Point Cook; Warrant Officer Alan Miller; Corporal Colin Bradmore, an historian in his own right; and finally Mr. Peter Coombs.

HIGH ADVENTURE

Contents

PUBLISHER'S NOTE

As all the photographs presented here are over sixty years old, it is inevitable that there should be some quality deficiencies in some cases. Many of the rarer shots had to be enlarged from tiny originals but have nevertheless been included as they make the historical record more complete.

SPECIAL NOTE

Until supplies run out, a limited number of the centre color pages of this book will be available direct from the Publishers as a separate item. If mounted, these prints make an eye-catching wall display of our own Australian-flown aircraft. For those who wish to purchase two (permitting all four pages to be displayed) there is a small price concession. Please send a stamped addressed envelope or International Reply Coupon for particulars. A proportionate number will be held in reserve for a short time as a service to our overseas customers.

The Publishers.

Prologue in Pictures

The first piece of equipment to be unloaded at Point Cook, Victoria, was a huge tent. It soon became a temporary hangar for aircraft of the newly formed Central Flying School.

Upper. The big tent erected and dispersed amongst the pines. Smaller tents were provided for the handful of officers and men. The aircraft on the edge of the trees was one of two BE2A two-seaters. The photograph was taken from the other BE2A. Port Phillip Bay was about two kilometres away to the right. Lower. Of the five aircraft comprising the machines of the Central Flying School, two were Deperdussins. One was an underpowered machine, used for taxying only. The other was flyable and is seen here on one of its few flights in Australia. It crashed on 2nd March 1914, only seven days after it first took to the air.

A BE2A coming in over the aerodrome boundary at Point Cook. An advanced type for its day, it was intended for use as an advanced trainer at the C.F.S. At the controls was Lieutenant Eric Harrison, one of two instructors at the school.

Upper. Lieutenant Harrison and Brigadier-General Gordon, the Area Commander, aboard a Bristol Boxkite elementary trainer. On this occasion during March 1914, a gusty north wind was blowing, but a worried Harrison successfully negotiated the general's first flight. Lower. Honorary Lieutenants Henry Petre (front) and Eric Harrison about to go aloft in a C.F.S. BE2A in mid-1914. Harrison, characteristically, wore his golf cap reversed.

The original C.F.S. hangars under construction on the edge of Port Phillip Bay. The Challingsworth engineered girders were hauled from Laverton rail siding, some nine kilometers away, by this fine team of draught horses.

Lower. A Bristol Boxkite in front of the same half-constructed hangar. This shot shows clearly the Boxkite's extended upper wing panels which could be removed to allow faster flight. Harrison, wearing a pilot's helmet, is in the instructor's seat. Note the waiting trainees in their Balaclavas.

If there is anything worse then a photograph of a Boxkite in front of a half-finished hangar it must be of a Boxkite in front of a Boxkite! This shot is important for two reasons; firstly it shows the fine type of hangar erected for the school, and secondly, the rear machine is actually a replica Boxkite made at the school around the middle of 1915.

Upper. Lieutenant G.P. Merz at the controls of a Bristol Boxkite. A C.F.S. graduate, he became the first Australian aviator lost in action, being killed by Arabs whilst serving in Mesopotamia with what was called the Half-Flight, Australian Flying Corps. This unit was commanded by Captain Henry Petre, previously of the C.F.S. instructional staff. Lower. Lieutenant Richard Williams in a BE2A prepares for a flight as Lieutenant T.D.W. Manwell (with white band around cap) waits to the right. Williams, later to become the first Air Marshal of the R.A.A.F., was the first Australian serviceman to solo at the Central Flying School.

By July 1915 not only was the Bristol Boxkite replica under construction, but facilities for officers and men were taking shape. This photograph shows the officers' quarters.

Some of the young officers who used the officers' quarters for the duration of their flying training. Many of them would soon feature in the war in the air. Left to right, seated on rail: Lieutenants F.H. McNamara (later V.C.), C.D. Merrett, E.G. Roberts, A. Murray Jones, Moseley, V. Hall, Captain W. Sheldon and Lieutenant A.W.L. Ellis. All were from the Third Course except for Hall and Sheldon. The man at the top right is unidentified.

Williams and Wackett with a BE2A. Lieutenant L.J. Wackett, a C.F.S. original, needs no introduction to those who study Australian aviation whilst Williams was to head the R.A.A.F. in its formative years.

Extra-curricular activity! Harrison is seen here in the seat with Captain Bill Sheldon on the protective anti-overturn pick. This was not normal practice, just a sample of unusually daring skylarking which was surprising in view of Harrison's ultra-conservative attitude.

Lieutenant Ray Galloway, Assistant Instructor to Eric Harrison, now a Captain. Galloway is seated in the non-flyable 'taxi' Deperdussin monoplane on which students gained their first realistic feel of aeroplane control surfaces.

Lieutenant A. Murray Jones, wearing his "Australian Military Forces" wings, photographed just before leaving for the Middle East. He was typical of the young airmen who were soon to be engaged in a new realm of conflict.

Officers of No. 1 Squadron, Australian Flying Corps, prior to departure for Egypt and operations against German and Turkish Forces. Some were later to play an important part in the air war on the Western Front. Top row: Lieutenants A.L. McNaughton, C.J. Brooke, A.D. Badgery, C.A. Kelly, S. Woodrow, A.E. Geere. Middle row: Lieutenant R. Ross, Captain W. Sheldon, Major A.A.J. Brown, Lieutenant-Colonel E.H. Reynolds, Lieutenants R. Williams, E. Roberts, R.S. Brown and J.L. Treloar. Bottom row: Lieutenants P.H. Newland, W.E. Hart and L.J. Wackett, Captains D.V.J. Blake and W.H. Anderson, Lieutenants F.H. McNamara and A. Murray Jones.

Something Different from the Infantry

Fresh from the Commonwealth Bank and the Brighton Rifles, young Lieutenant Arthur Henry Cobby ready for the Central Flying School, Point Cook, and the Sixth Course. Within two years he would attain the D.S.O. and D.F.C. with two bars, and shoot down 24 enemy aeroplanes, with a further five unconfirmed, and account for 13 German field balloons.

CHAPTER 1

A party of some thirty officers in various forms of military dress emerged from the train at a little wayside railway station at Laverton, Victoria, late in 1916. One or two were acquainted, but generally we were strangers to each other, and for various reasons we comprised a course for instruction as pilots at the Military Flying School at Point Cook. Nobody other than ourselves seemed interested in that. The school was six miles away on the shores of Port Phillip Bay, but it seemed that no one had thought to provide the means of transport, so some of us walked the six odd miles and arrived very dusty and tired in the late evening.

Although I had made many previous attempts to get away to the War, I had not been permitted to do so. I was a clerk in the Commonwealth Bank, which was handling a tremendous amount of war financial business, such as War Loans and at the same time was deeply involved in a large scale expansion. So I was privileged to consider myself reserved for service at home. This was never acceptable to me, however, and not because I particularly wished to enjoy the excitement of war, but because for many years I had been a Militia officer. All my earlier associates in the Regiment had enlisted and most of my friends, and I was still a Saturday afternoon soldier at home. Naturally I was always conscious of the fact, and was made more so by the covert suggestions of patriotic but blundering "flag-waggers", who frequently alluded to the fact that "Harry Cobby who had been lording it over the local lads for so long, is still walking around in uniform of a Saturday" and so on. And, of course, explanations carried no weight whatsoever. The mere fact of looking healthy and of military age was quite sufficient for the flagwagging enthusiasts. There was only one place for anyone with those qualities, and that was at the War. I still hold a certificate of exemption from the War — thumb prints and all. Some two years of this was more than I could stand, so finally I put the hard word on the Governor of the Bank, pointing out that I had the option of resigning, or jumping under a train if I was not permitted to join up. He saw the reasonableness of my request, especially as I had based it on my getting into the Flying Corps.

There were two reasons why I selected the Flying Corps. The first that two of my friends, brother officers in the Brighton Rifles, had already completed a course at Point Cook and had gone overseas with the Australian Flying Corps. They were Frank McNamara (later V.C.) and Murray-Jones. That in itself meant nothing other than suggesting the idea of serving in something different from the Infantry. The other reason was more subtle. I felt that being a bank clerk, the Governor would be of the opinion that I had almost no chance of being selected, especially as a fairly high technical knowledge was required. I was careful to explain all this, but I was also careful not to mention that I had practically paved the way to selection by telling my hard luck story in the right quarters in military circles. So finally I received the necessary assent if I was fortunate enough to be selected. I was back in an hour with everything signed on the dotted line. I know quite well that my decision to join the Flying Corps was not based on any strong desire to fly.

"One sat in a very exposed position." Cobby was in the seat with 'Jonah' looking on.

We reported to the Adjutant at Point Cook, who knew nothing whatever about us, and found that like many other similar arrangements of those intense war days, notice of our course had not been passed on to the people who had to take us in hand. Worse still, the previous course was still in training, and the accommodation was only sufficient for one batch of thirty at a time. So the obvious action of double banking was taken — all the old course in one wing and ours in the other. So started my flying career.

I roomed with a nuggety chap of very bright disposition, one. O.J. Jones, from Tasmania, and we chummed up particularly well. Opposite to us was Tommy Hemson, with whom I had worked in the bank some years before. His bunk mate was Billy King, a law graduate from Sydney. We four teamed up and through various moves of the squadron remained together for many months. Jonah had a mass of fair frizzy hair of a particularly tough variety and one of his habits at a particular stage of conviviality

Left. "A mass of struts and booms and wire." Harrison at the stick.
Lower. "... what made the wheels go round was hammered into us as hard as we could take it." Ground instruction on an engineless Bleriot monoplane.

"We stood around most of that particular period." 'Jonah' took the picture; 'Freeda' and Cobby are second and third from the left respectively.

"... an establishment that contained but two instructors and one assistant instructor ..." Captain Eric Harrison, chief flying instructor, standing. Lieutenant Ray Galloway assistant instructor, seated on wing, above right.

was to lay on his back on the floor and demand that we pull him up on to his feet by that same hair. He was a ring-leader in any devilment that took place. Hemson was of more thoughtful disposition than the rest of us, but Billy King was possessed with a lively and clever satire.

The dearth of experience or knowledge of Military authorities in those days in the teaching of the technique of flying can be well understood. Even so, to wish thirty raw aspirants for flying knowledge onto an establishment that contained but two instructors, and one assistant instructor and three aeroplanes, is difficult to understand. But to superimpose this thirty on top of another thirty that were under instruction, was even then entirely beyond comprehension. Apparently Australia was committed to send a certain number of squadrons overseas and it had been arranged that their training would be completed somewhere "on the other side". So after a few days the other course packed up, irrespective of the amount of training they had had, formed up in camp locally with their mechanics and other staff and departed for somewhere overseas as a Squadron.

Then we started to be taught the art of getting a contraption of wires and sticks off the ground and back again as near in its original state as possible. The theoretical side of the business, aerodynamics (a ghastly subject to one stale on mathematics), engines, and what made the wheels go round, was hammered into us as hard as we could take it. The instruction was part of the job of the flying instructors, and I still cannot imagine how they survived the day. Class lectures would take four or five hours per day and they would be giving dual instruction in the air for somewhat similar periods. This meant 5 a.m. on the tarmac, and eight or nine o'clock to cover the ordinary day's work. With thirty of us to be handled, it also meant that although we were allotted to flying parties for flying instruction, we stood around most of that particular period.

The greatest care was always exercised by the people in charge to ensure that the aircraft were not exposed to undue risk of being damaged. The state of the weather was of utmost importance, and the instructor's silk handkerchief was usually employed to find out if there was any movement of air. Failing this, a wet finger held up was

"... the accommodation was only sufficient for one batch of thirty at a time." A rare shot of the tents at Point Cook.

"You sat directly behind the instructor with your feet on a couple of rests." Galloway, front, prepares to instruct a pupil.

used to determine whether the machines could be wheeled out of the hangars or not. And, of course, the machines were something that one only sees nowadays in museums, or reproduced in photographs with the pilots wearing caps of sporty appearance back to front — supported by long black moustaches. Yet they performed their functions well. It was difficult to keep the engines going, as they were not by any means reliable in their component parts, but every mechanic had odd springs and pieces in his pocket, and delays in getting away, though frequent, did not last very long. The aeroplanes were Graham-White and Bristol Boxkites, a mass of struts and booms and wire, upon which one sat in a very exposed forward position with nothing around one at all. Pants were tucked into socks, the cap (at

"... of course the machines were something that one only sees in a museum nowadays." From Cobby's own album this photograph shows a typical scene at that time. From left to right, a Graham White Boxkite, a Bristol Boxkite, and another Graham White Boxkite which is partially obscured by a Maurice Farman Shorthorn.

"The instruction was part of the job of the flying instructors, and I still cannot imagine how they survived the day." A Maurice Farman Shorthorn awaiting rescue after a mishap.

that time made of thin leather and just like a pillow slip with a hole for the face and a cord round the neck) and motor bike goggles, and you were all set. You sat directly behind the instructor with your feet on a couple of rests, and your legs alongside the teacher's and nothing to hang on to. You just sat up in space. A vertical control stick on the right, and after ten minutes of floating around the air, with one or two sentences bellowed at you by the instructor, which you could not understand, your instruction was finished until the next time — perhaps a week later. The only two things that impressed themselves on me during the period, was to keep the front elevator on the sky line and to watch the oil drip feed. I presumed there were many more things to learn, but they did not come my way.

The structure of the buses was simple. The engine — a 50 h.p. Gnome, was behind and pushed you along. A pair of wings from which two sets of converging struts met out well in front to support the elevator, and two larger ones called booms, projected aft to support the tailplane and rudder. Piano wires braced everything in all directions and the old air force jest that the proof of correct rigging consisted of releasing a canary within the wires and if it got out the rigging was out of plumb, was capable of being believed — almost. The machine sustained flight at about 40 m.p.h. Of course the type of aircraft was the only sort available at that time. A more modern and comparatively high speed job was nearing completion of erection and one or two of us were fortunate enough to have a trip in it. This was the Morris Farman "Shorthorn", so designated to distinguish it as a type from the "Longhorn". This "Shorthorn" had the elevator in the rear, hence the title as there was no forward projection. The Longhorn had it out in the front like the Boxkites. The Farman, to our minds, was a marvellous advance in type. It had a small bull-nosed box in front of the wings, called the nacelle, for pilot and pupil and one did not have to adopt the cyclist's habit

A BE2A of the Central Flying School bearing the designation C.F.S.2 and showing signs of wear and tear. The outsize wheel on the port side was a makeshift replacement from a motor cycle. See color illustration on Page 53.

"There was another 'real' aeroplane in the hangar ... a single-seat Bristol Scout ... and we were wont to gaze at it in awe ..."

of tucking the trousers into the socks. You could descend from it looking something like an officer and a gentleman, plus his cuff-bottomed "trous". Otherwise, the bus was still a Boxkite, a bit larger and heavier certainly and with a stationary engine that pushed you along with far more reliability. It had still more wires to the square foot, but an advance in speed of over 20 miles per hour. There was another "real" aeroplane in the hangar that was never allowed out. It was too dangerous. It had a speed of nearly 80 miles per hour and was a single-seat Bristol Scout. To we novices it seemed the last word in high speed aeroplane construction, and we were wont to gaze at it with awe as it rested behind its wired off enclosure. But we discovered a few months later that it was one of the simplest machines to fly in existence, in fact when we arrived in England, we saw dozens of them at one 'drome being tossed around the sky by youngsters with only a few hours' flying experience.

Six weeks after our arrival at Point Cook, we were given the same treatment as our predecessors. We were told to pack up and go into camp under canvas, form into a squadron and proceed overseas. Our flying times were negligible, but we had gained some idea of what was before us, and how little knowledge we possessed to meet it. That is something of course — to know one's de-

Cobby, 'Freeda' and some tough looking friends on Port Melbourne pier, about to embark the Orient Line ship *Omrah*.

ficiencies is always of value. So twenty-five of us became embryo pilots and one step nearer to the war was negotiated.

A few days later we were at sea on board the Orient Line ship the "Omrah", our destination unknown. We touched at Adelaide, Fremantle, Durban, Capetown, Sierra Leone, and Plymouth, taking nearly twelve weeks to make the trip, but it was a very pleasant one. Some thirty midshipmen from the Naval College at Jervis Bay were on board bound for the Old Country, and the A.I.F. was represented by a tunnelling company.

Our approach to Plymouth, although we did not know where we were until we actually got ashore, was more impressive by virtue of the tremendous amount of caution imposed upon our last couple of days at sea. The ship, of course, was completely blacked out at night and escorted very closely by destroyers during daylight, and the last night a fast run was made until we were apparently coming up close to the port. Then at dead slow we moved into berth without a light or an indication of any sort as far as we could see, to show the way. After doing a freeze for an hour or two, we disembarked and made an all night move by train to a temporary camp at Perham Downs on Salisbury Plains, within sight of the Central Flying School at Upavon. Snow several feet thick was everywhere and the cold bitingly intense, and we chaps were mostly unprepared as far as clothing was concerned, to combat it. We could not get warm and we piled on every piece of clothing we could. The nights were frightful until we "found" some kerosene heaters, and although the smell and the "fug" were terrific, we did get the temperature of

Upper right. The unit soon to be known as No. 4 Squadron, Australian Flying Corps, embarked on 25th October 1916 under the command of Lieutenant Andrew Lang who is shown here. For a short time he had been the No. 2 Instructor of the C.F.S. Right. Six young men on their way to help the mother country. 'Jonah' was on the left, 'Freeda' was holding the lifebelt and Cobby was immediately behind him. The rest are unidentified.

Lower left. Cobby's own caption for this photograph was: "Me as I fly at Harlaxton." Lower right. 'Jonah' at left, Stafford (?) in the middle, and Cobby, during their six weeks study course at Reading University.

our huts up sufficiently to be able to sleep. A fortnight of this with nothing to do, and we moved to Castle Bromwich aerodrome just outside of Birmingham.

We received our first impressions then of what aeroplanes could do. As we moved into the station, which was quite close to the 'drome, the air was full of machines and everyone seemed to be able to handle them with the utmost nonchalance. To our inexperienced eyes it was rather marvellous, even though some of our old friends the "Shorthorn" were chugging around. Most of the machines were Sopwith Pups and Avros, and it immediately became the ambition of all of us to fly those natty little Pups.

The town of Birmingham became our regular haven during leisure hours when the pay book would stand the visit. I have often wondered what the hard-bitten Midlanders we used to meet in the Midland Hotel thought of us. The basement bar in the Billiard Room was a meeting place of an evening of a number of factory managers from Vickers and other large munitions of war concerns, and they were very friendly towards us Australians. We were under the impression that we created admiration for ourselves by our manner of drinking. We were new to the partaking of strong waters and like most beginners started the habit for company's sake by taking short drinks like liqueurs or port wine. We always noted that a shudder went round the circle every time we did so and had half an idea that we were being regarded as something super. It was some time before we realised what a severe shock it must have been to well seasoned whisky drinkers to have to watch us take our thimblefuls of Creme de Menthe or Cherry Brandy. No wonder we staggered them.

However, we were destined not to remain long in this area. As soon as we had settled down, the mechanics were split up into small parties and sent to various places for technical instruction and the "would-be" pilots were treated somewhat similarly, and batches of us were sent to various Flying Training Stations, to get on with the business of learning to fly. Three of us, Tommy Hewson, O.J. Jones and I, went to Waddington in Lincoln and started off on our old friend the "Shorthorn", known in England we now discovered, as the "Rumpty". A couple of others went to the same 'drome but to another squadron, and we commenced serious work.

One of our chaps named Freedmann, who was with the other squadron, was waved off for solo and as was customary when anyone went off alone for the first time, we turned out to watch the result. "Freeda" got off all right and managed the circuit of the aerodrome quite well, but in coming in to land, he forgot to shut off his engine and of course the old "Rumpty", for her type, developed an enormous speed. When he pulled his stick back to land, his speed zoomed him up in the air about a hundred feet, so he shoved his control forward and shot at the ground again and once more he zoomed up. This went on right across the aerodrome, for all the world as though he was on a switchback railway. When just level with us he managed to hold the machine flat, still with his engine full on and as he disappeared behind a hangar on the edge of the 'drome there was a loud crash, then a series of further bangs and we rushed to the roadway to see what had happened. All we could see was a twin pair of wheels spinning in the roadway and nothing else. "Freeda" had smacked a ditch, then gone through a hedge, over another ditch, and then what was left of the "Rumpty" skidded about fifty yards into the field. The farthest forward portion was the nacelle which was then resting on the ground with a dazed Freeda standing up looking back at the pieces of his aeroplane strewn behind him. His color was a shade of green, but he was unhurt. Everything else, other than himself, had absorbed some of the various impacts. Then there was a roar of a motor cycle as his instructor dashed onto the scene and he started to tear into poor old Freeda like a fishwife (why a fishwife, I don't know, it is an accepted description, but I should imagine any other sort of wife would do as well under provocation.) Up till then Freeda's mind must have been concerned with the fact that he had had a remarkable escape, why he did not know and the fact of having done something wrong in his flying, never occurred to him. But to be suddenly jumped upon by his instructor for daring to bust up one of his machines, when he was prepared to be killed learning to fly or by the enemy if necessary, seemed entirely out of proportion and un-

"... there was a loud crash and a series of further bangs." The complete demolition of a Maurice Farman Shorthorn.

Cobby's own caption reads: "This is me in the first Mono-Avro I flew and in which I looped five times. She was smashed up next flight by another chap." Avro 504As and Js comprised "A" Flight in No. 68's line-up when being used for advanced training prior to the squadron's mobilisation.

reasonable. And then the balmy air of Lincoln was subjected to a flow of language in good Australian, that got stronger and stronger as his senses came back under control and his feelings of indignation and injustice asserted themselves more clearly on his woozy mind. After a serious crash, here he was still alive and prepared to give it another go and to be ticked off for bending an aeroplane, was just too much. Freeda won, unfortunately, for a little later on he piled himself up fatally.

Hewson, Jonah and I moved round to several aerodromes during our training, from *ab initio* types to something more advanced, just an hour or two in the air at one place and then a change to somewhere else. Being Australians, practically foreigners in the eyes of most Royal Flying Corps people, we were not only given some latitude, but very few of the units to which we were posted were interested in us. They treated us extraordinarily well, but were generally of the opinion that Australians were not true to type unless they broke all the usual ordinances and regulations, and we found it was sometimes a difficult character to sustain. I know that Jonah and I found it very easy to pull the leg of a young Padre, who took us three uncivilised colonials more or less under his wing, at one station. It started when he discovered Jonah and I playing billiards in the mess and it was just too easy. He wanted to know if we had billiard tables in Australia (of course it might have been the standard of our play that suggested the thought to him), and we informed him we had never seen one until we got to England. The conversation went from billiards to churches, and our joking remarks that there were practically no churches in Australia but only "pubs", was taken seriously, and his great thought was that it would be a wonderful life work for someone to be able to place at least one church in every town in Australia. Later when we tried to assure him that there were thousands of churches in the country, he just shook a finger at us and said, "I have heard what jokers you Australians are" and would not believe us.

So the stories we told of the dangerous kangaroos and the savage "bush" goats, would have shocked the home folk. We told him of how the male kangaroo's bite was one of the most deadly in the world, that they would sneak into our "lean-to" houses at night and bite the unsuspecting slumberer, who would be found dead in his blankets in the morning. How the wild goats would visit the towns at night and chew down everything of wood they could get their teeth into, fences, telegraph poles, and so on — that all communications of some towns would be cut off for days when a mob of goats descended on them from the hills. These goats were as big as a pony, having crossed with the dingo — the wild wolf of Australia. But when after a week or two we discovered he was a particularly nice chap, although unsophisticated, and tried to retract what we had said he would still laugh and say, "I know you wild Australians." We could only be taken seriously when we were drawing the long bow; when we became serious and truthful, he was politely sceptical. The difference between our tall stories and the truth, was apparently so marked that it appeared too difficult to accept the truth.

After moving from place to place for a month or two, we were instructed to report to No. 68 (later No. 2) Squadron, Australian Flying Corps, at Harlaxton near Grantham. This moving round was pleasant and interesting to us, but it did not react to our advantage as prospective pilots. As soon as word was received of our posting to another station, we immediately applied for 48 hours for the last two days, in order to fix up private business. This was always granted, then we would report to the new school and make a similar application, with like results. In consequence we saw a good deal of London. But Harlaxton was different. We were duly shown in to the Commanding Officer, a well-known man in Australia — Colonel Oswald Watt. He had been in France when war broke out, and although somewhere in his forties, had immediately joined the French Flying Corps. He did a lot of flying with the French in the early months of the War, and then switched over to the Royal Flying Corps. He became a Flight Commander in No. 1 Australian Squadron in Egypt, and then Commanding Officer of No. 68, which was then a training unit, and he knew the habits and instincts of his

"My first crash, getting out of a rough field when the engine conked, and we lobbed on a hedge."

fellow Australians. As soon as we saluted, he demanded to know where we had been, and we told him truthfully "on 48 hours leave from so and so" and proceeded to show our passes. However, this was not what he was talking about. He was referring to the fact that the war had been going on for a couple of years or so and he wanted to know why we had not plucked up courage before this to do something to help win the war. He dressed us down like a trio of pick-pockets and laid it on thick and heavy. It was rather extra-ordinary to be jumped on like that, without waiting or asking for any explanation. In any case we were still only kids, although we thought we had assumed men's estate and resented his attitude. Tommy Hewson found his voice first and said he was astounded at such an attitude of mind from an Australian, particularly when every English person we had met had expressed themselves to the effect that it was very decent of the Dominions to hop in so thoroughly. We all had something to say, but not much as "Ossie" just silenced us and instructed the Adjutant, Roy Phillips, to march us out.

We discovered from some of the other chaps on the 'drome, who had known him for some time, that he was terribly keen on beating up the enemy and that he was absolutely intolerant of everyone not in the War in some capacity or other. As a matter of fact we agreed with this, but felt that a little reasonableness would not have been out of place. Everyone said he was a particularly fine man, and an excellent Commanding Officer, and as nothing could be done about it, we just put up with it. Later we found him to be one of the kindest and most thoughtful people we had ever met and this feeling remained. Obviously we did not try and work the forty-eight hour trick on him.

There was a vast difference in the manner in which we were treated at Harlaxton, to that meted out to us at Royal

"We were duly shown in to the Commanding Officer, a well known man in Australia, Major Oswald Watt." Ossie Watt is shown here as a Lieutenant-Colonel. This picture was actually taken some two years after the time Cobby describes.

Flying Corps schools. We were Australians with Australians and no longer gentlemen visitors from the Antipodes and instruction and comment was direct and to the point. The senior instructors were Guilfoyle, Muir and Matthews and they just put us through the hoops. We were started off on Avros, something in our experience that was more like a proper aeroplane than anything we had touched. There were no frills about the place at all. A few hours with an instructor, a few "straights", that is a fast taxi and then lift the machine off and put it down again, then one dual flight right away, and you went "solo." Not only did we have to follow a fixed sequence of movements in the air, in sight of our instructors on the ground, but we were all under a moral obligation to loop the loop on our first solo. Somebody had started the habit some little time before and although nothing was ever said until after you had done it, even amongst the pupils, it had to be gone through otherwise you lost caste. It did not matter how many times you looped the loop later on, it had to be done on the first solo. It must seem strange to pupils these days under training, to imagine that anyone would be nervous of looping an aeroplane. It is, of course, about the simplest evolution to perform, but we were all timid of carrying it out at that period of our flying history.

I had already gone solo on a "Rumpty", but you didn't throw those around, you just remained sedate, crouched over the controls with your feet on the rudder treadles, as though playing a harmonium. But the Avro was a different proposition, it would stand almost anything. So I went solo, and all my friends stopped to see the event. Three or four times around the aerodrome to comply with instructions, were to be negotiated first, then the custom was to move to a position just away from the 'drome where you could be seen, and then loop. It is strange how many small things can exist that help you to stall off doing something you are not very keen on. First, when just about in position, I thought it better to be dead into wind, so I turned back and returned on a slightly different course. Then when just about to put my nose down, I thought I would ensure that the "blip" switch was working properly, but after about four runs up and down, I realized it had to be done, so down I put the nose until the speed showed about 160 miles per hour, then easing the stick back I zoomed up and then while keeping the stick back, the machine went right over and just as the ground came back into sight I throttled the engine. A few feet more and I had looped the loop. It was just too easy so I did another one, and then returned to the 'drome and managed a fairly decent landing. Nothing was said by anyone, they had all disappeared except my instructor. But I felt like a million dollars, and as I have no idea how a million dollars feel, the description is apt. I did feel pretty good though.

I had to go to hospital for a week or two just as I qualified for my wings as the result of a minor crash. I had force landed in a field alongside the aerodrome owing to the high tension ignition wires on the engine coming adrift, and had made it O.K. The area was very small, about three hundred yards across on the longest run, and was surrounded by a built up stone wall about four feet high. Matthews asked if I could manage to get out all right after they had fixed up the engine and I was confident that I could, so getting into the far corner I started off. Alas for the egotism of youth — I did not get right out, only the front end managed to. Just as I got off the ground the engine spluttered and I hit the wall with my undercarriage

Another Sopwith Two-Seater. No. 68's "B" Flight aircraft consisted of four Two-Seaters and two Sopwith Scouts (Pups). Cobby is in the front cockpit of this one and the others are unidentified.

Graduates pose after the six weeks course on Military Aeronautics. Cobby did his at Reading University; the other centre was Oxford University where this picture was taken. Seated second from the left is Lieutenant E.J. 'Tabby' Pflaum; second from the right, standing, is Lieutenant R. King. Both were to play a major part in Cobby's life as a fighter pilot. Next to King on the end is Lieutenant G. Jones — later Air Marshal Sir George Jones, the writer of the foreword of this book and himself a No. 4 Squadron fighter ace with six confirmed victories.

and broke the machine in two, with the tail inside the field and the nose sticking in the ground outside. A piece of broken longeron, which is one of the fore and aft members of the fuselage, stuck into my thigh and made a nasty gash. It was not very serious but it became infected and I had to stay in hospital for nearly three weeks. When I came out Hewson had gone to France and Jonah had been posted elsewhere. I never saw Tommy again as he was shot down within a few days of going overseas. It was a sad business. After some eight or nine months of being together in the closest of friendships, it was a shock to realise he was gone. It also brought to mind that there *was* a serious side of the war.

A Sopwith Two-Seater of No. 68 (Australian) Squadron, Royal Flying Corps. No. 68 (later No. 2 A.F.C.) was awaiting equipment and operating as a training squadron for the R.F.C. Note the kangaroo on the nose. It is almost certain that this type of decoration was originated by Major Oswald Watt the O.C., who sported Australian emblems on his Farman biplane when he was flying with the French Aviation Militaire.

An unpainted Sopwith Two-Seater being overhauled. Note the kangaroo on the fuselage. This type, also known as the 1½ Strutter, was a front-line aircraft relegated to training duties. It had an 110 h.p. Clerget rotary radial engine and was a handy intermediate link between Avro 504s and fighter types.

I switched about from one type of machine to another for a few weeks and then was posted to the Central Flying School at Upavon to fly Camels. Before I left we had a night in Grantham, but money was too scarce to hire a car to make the return journey and to still have enough to spend, so one of the number thought of an excellent idea. The aerodrome was still more or less under construction, and a tractor was used to haul a road grader on its job. The bright youth had the thought to drive the tractor in, but as only about two people could get on the tractor, we took the grader along as a trailler and made the journey with the minimum of expense — and comfort. It was regrettable that the old man, 'Ossie' Watt should be coming back by car at the same time as we were, as it cost all the other chaps leave for a couple of weeks, but I was lucky in that respect as I was posted from the next day.

The situation of the Central Flying School at Upavon was really pleasant. It was on the Salisbury Plains, near to several pretty villages, all old fashioned and interesting. The downs themselves were delightful for walking on, as well as flying over. The School itself was the official father of all Military flying in England, and although it was made up of old and shabby hangars and the usual temporary buildings, it possessed many attractions. There was little history to military flying in those days, but whatever there was had seen its infant nurture in this school, and consequently the standard of instruction was high, at least for the times. A chap named Longton was responsible for

Another heavy arrival during training, this time by an Avro at Harlaxton in June 1917. The glossy black engine cowling is very noticeable here.

A little Sopwith Scout looking rather sorry for itself. Despite its manoeuvrability, some Australian instructors were of the opinion that its undercarriage was not really robust enough for the hurly-burly of training.

my tuition, and he was excellent. The manner in which he threw a Camel around was astonishing and after several flights on a Sopwith Pup, I was sent off in the rather frightening Camel. This machine had a bad name as so many fellows, even experienced pilots, managed to get piled up in them. And you could not be trained by anyone else in their tricks, only by word of mouth on the ground, as they were single-seaters. Longton, though, was a painstaking fellow and he devoted himself to drilling into my head all the things I should know about them. He at least inspired some confidence in their flying capabilities. So I duly went off in one, and the experience was strange. It climbed unusually fast without my help. Not only that but I seemed to be going straight ahead and the ground passed by slightly sideways, until some little time later the aerodrome appeared in front of me again, so I glided in and landed. Longton told me I just went slowly round in a large circle in a flat turn, and he also corrected the foot pressure fault that caused it. Next time I managed better and from then on I improved considerably.

I received my pilot's graduation ticket soon after arriving at C.F.S. I had qualified some month or two previously, but when I went to hospital at Harlaxton this had apparently been mislaid, and the C.O. at Upavon suggested I went through the tests on the spot so as to avoid further delay. This was disappointing at the time as I lost a fair amount of seniority. But it did not matter much in the long run and it was with some pride that I stitched on my wings. It was the culmination of ambition and the chest could not stand out far enough to satisfy my feelings.

Warning then came that I was to return to Castle Bromwich where No. 4 Squadron was going to go on to a war footing in preparation for France, so back I went, and found that practically none of the original pilots had been posted to the Squadron. They were nearly all new, and in October we switched over as a Service unit.

Just after reporting back to Castle Bromwich word was passed round that Jonah had piled himself up badly and that he was in Wandsworth Hospital seriously injured. I obtained leave and went down to see him. He was swathed in bandages and splints and although badly knocked about he was going to be all right except for one serious piece of damage. When he hit the ground he had displaced his heart in some way or other, and was not going to be allowed to fly any more. He was terribly upset and depressed, but picked up some weeks later. He told me that he had seen Billy King and he also had been washed out from flying. Billy had had several bad attacks of asthma and on several flights had passed out more or less in the air. So Billy too was not to reach the war. I went back to Birmingham almost as dejected as Jonah.

We immediately commenced to collect our war equipment — aeroplanes, transport and so on and our main efforts were devoted to seeing that we found out all about the various items with which our machines were equipped. Guns and gun gear received special attention.

I had not done any air gunnery at all, but was advised that a course would be arranged — but it did not eventuate.

Upper. The end of Captain Stan Muir, "A" Flight Commander of No. 68 Squadron, near Harlaxton. His death even before Australian units were involved in action emphasised the seriousness of military flying. The machine in this instance was a DH5 fighter, the mobilization equipment of the squadron. Lower. Taken at Central Flying School, Upavon, this photograph records Cobby's first encounter with the renowned and sometimes reviled Sopwith Camel F.1. Although No. 68 had dealt with higher training, a pilot's graduation did not take place until a "finishing school" had been completed for service qualifications. The next step in training was active duty.

Novices

CHAPTER 2

December 18th 1917 was the day we moved off from Castle Bromwich. I accompanied the ground personnel, mechanics, etc. and transport, and other than that we were bound for France we had no knowledge of our destination. The night was perfectly still when we left by ship, and naturally there was not a suggestion of light showing anywhere. It was freezing cold and the air almost had a crackle in it as we silently left port. Everyone was keyed up to concert pitch. Of course we stayed up to try and see what was happening, but all we knew was that we were slowly moving through a very still sea. After some time, a purple light appeared close alongside, swaying in the blackness surrounding us, and a voice distorted by megaphone came out of the darkness carrying some instructions to our ship. Our transport was only a small affair, and only just accomodated the 250 men on board. Then the light moved right up in front of us, and for another hour we followed it through the night and it was the only indication that someone knew where we were going. We zigzagged several times, and then when dawn just gave sufficient light to discern objects close at hand, we found ourselves in a harbor and moving slowly up to a stone jetty. It was Le Havre.

Motor lorries, tenders, cárs, and workshop lorries were formed up at a small concentration camp, and then we found out we were on our way to Bruay. The road was frozen, the vehicles new and not tuned up properly, and they gave a lot of trouble. The heavy lorries with solid rubber tyres without chains, were particularly troublesome on the glasslike icy roads. They just refused to go up hills, so light tenders had to carry out a lot of haulage. Even so, several of them skidded backwards into ditches. Finally we made our destination.

The people in the air also had a lot of trouble. Their equipment was also new, and engine trouble put a lot of them on the ground, and it was several days before everyone was on the spot together.

The first sight of our new location was most uninspiring, in fact it was depressing. The Bessoneaux hangars, contrivances of timber and canvas, and our quarters, looked more like a gypsy camp than anything I could imagine. We were lucky in that there was another squadron, No. 40, on the same 'drome and they looked after us, as least as far as the most important part of existence was concerned, namely, the provision of meals.

I can remember the nervousness that assailed me during the months of training in England, when I gave thought to the fact that as soon as I was qualified to fly an aeroplane, or perhaps sooner, I would be sent off to the War to do battle with the enemy in the sky and on the ground. I quite freely admit that if anything could have been done by me to delay that hour, I would have left nothing undone to bring it about, but unfortunately, or perhaps fortunately, I had no control over the supply and demand of pilots. Much had been written and spoken of the nerve and bravery of the chaps in the air during the previous days of the Great War, but I felt that in many instances it was born of desperation, in others an outburst of temper or rage, in that the man concerned was placed in a position of being shot at without a great deal of oppor-

"Our machines were a type that possessed a very low cunning, but once you became used to their little idiosyncrasies, were a pleasure to fly." By the time this photograph was taken at Bruay in March 1918, most could give a good account of themselves in their Sopwith Camels. On the extreme left is the O.C., Major W.A. McClaughry and second from the right is Cobby.

tunity to hit back. This, acting like the red rag to the bull, so aroused the ire of fellows, that they went beserk against all sorts of odds and were acclaimed brave men. Admittedly they retained a sufficient measure of control over their physical and mental reactions to enable them to vent their spleen in the right direction. After a while they became used to the atmosphere and it more or less became a personal characteristic. Perhaps it was the same thing, but I am quite sure that of the twenty or thirty pilots who accompanied No. 4 Squadron to France, there were very few who could be regarded in cold blood as being naturally cut out for the role of hero. On the other hand, most of them would be able to do the Horatius act per medium of their "bus" and "Vickers", if they came up against fearful odds, but they were definitely quiet fellows when they first saw France.

I know full well my feelings of smallness and trepidation on the occasion of arriving for the first time in the theatre of war to try and prove my ability to bear a share of the burden in the air. I felt that I would never achieve the jaunty outlook on life that seemed so characteristically part of the make-up of the pilot who had been overseas. I little knew that quite a large proportion of his perkiness was due to the fact that whilst in England the enemy was not throwing lead and heavy shells at him.

We had arrived as a completely new fighting squadron of the Australian Flying Corps, our title No. 71 Squadron, Australian Flying Corps, but this was soon changed to No. 4 Squadron A.F.C., and our machines were a type that possessed many inherent tricks which they played upon the unwary. They seemed to possess a very low cunning, but were a type that, once you became used to their little idiosyncrasies, were a pleasure to fly. We came from all walks of life — farmer, bank clerk, college graduate, and one or two university people with degrees. One was a dentist, another was an engineer and another a doctor, but one and all I think were somewhat apprehensive at finding themselves after all too little training, about to come to grips with the enemy. Hearsay and a morbid curiosity had filled our minds with rather an unhealthy opinion that the German was far and away a better equipped and a more efficient pilot than the average Britisher of the time. This was an opinion that was induced, perhaps unconsciously, by the daily publications in the press of the latest victories of the outstanding German aces, such as the Baron von Richthofen, and the never stale gossip in the messes in the old country, of the dogfights from which the enemy always seemed to emerge victoriously. We did not then realize that quite a large number of the instructors with whom we had come in contact, had seen very little fighting for quite a number of reasons. The Air Force had developed so quickly and was still expanding at a fast rate. Almost any pilot who had returned from overseas, irrespective of the cause, whether from incompetence, disability or for a rest from the War, was put on to instructing, and a large number of the more advanced pupils themselves were given similar duties. Consequently the general knowledge of air tactics and War fighting among instructors was rather low and more academic than practical, and we unconsciously became saturated with a distinct feeling of inferiority. It was a state of affairs for which there was no remedy, as the successful prosecution of the War demanded a continual supply of pilots, but we were sufficiently long in the tooth to realize that the art of air warfare, as far as we were concerned, still had to be learned.

In the air a pilot's job is a strangely isolated one. Although formation flying and fighting demanded a very close cohesion among the members of the flight, in the early days one felt very much alone and at the mercy of chance. Lack of means of communication with one's fellows was, I think, the principal cause of this. You could not communicate with the other chaps about what was going on. Every possible eventuality had to be studied before leaving the ground. Plans had to be made to meet each contingency that might arise, and then your memory and a thorough understanding of the characteristics of your leader and your comrades had to be relied upon to dispose of each bit of trouble as it presented itself. There was no means of talking over what you were going to do; once something started you just went ahead and did it and the firm reliance one placed on one's comrades later on was the result of long and continuous association in the air under all sorts of conditions. In those early days we lacked this strong background of mutual support and understanding.

Our Commanding Officer and the flight commanders had some previous experience, but it was not vey extensive. McClaughrey the C.O., had done some little time on night flying Camels. Hughes-Chamberlain and Arthur O'Hara Wood had been with the Royal Flying Corps for some period before being attached to us as acting flight commanders, whilst G.C. Matthews, our Flight O.C. and my earlier instructor, had been overseas with No. 68 Squadron, A.F.C., a month or two before and had then been posted to us to make up our complement of flight commanders. The remainder of us were brand new and some of us had not even fired a gun from the air.

My total flying time, both instructional and solo, was about twelve or thirteen hours, and this on six or seven different sorts of aircraft. I think one or two of the others were even worse off, and here we were about to give battle in the air to the enemy as fighting pilots. This lack of experience would not have been so badly felt if, for instance, we had been attached to well established units where we could have been mothered for a while, until we found our "air" feet or wings. Several of the chaps, unfortunately, qualified for the heavenly variety only too soon. We also knew the fate of the mythical Icarus who stole the wings which his father Daedalus, had fabricated of wax. He perished, either because he flew too close to the sun and the wax melted, or because they were not strong enough to bear his weight. We felt something akin to friend Icarus, only our names would not be immortalised as his. Some attempt certainly had been made to give prospective pilots of our new Australian squadrons a brief term in France before forming up, but they were either shot down, invalided home, or were not available for some reason or other when we moved off. We were novices almost to a man.

Our first few days were spent in settling into quarters. There was little of interest in this, other than that of ridding our Nissen huts of the vermin that had taken possession of them. The whole country was under snow and I think everything that could crawl or hop had taken advantage of the shelter and had devoted its time to the multiplication of its species, and their fecundity must have been remarkable. Petrol and kerosene and plenty of bonfires gradually accounted for them, but it was some weeks before the blankets stopped moving under one.

Our aerodrome was on the outskirts of Bruay, a small town in the Lens coalfield district near Bethune, and that just about completes its description, and it was to be our home for some months to come. It possessed none of the amenities or facilities usually associated with a French town. It did, however, possess a most efficient set of hot water showers at one of the mines, and once one became used to bathing in the company of coal begrimed people of both sexes — the privilege proved a boon. If required, a day was by arrangement set aside for Flying Corps personnel to perform their ablutions without being subjected to the indifferent stares of the blase mine operatives, both *hommes* and *femmes*, but this was not always convenient. I for one followed the maxim "to do in Rome as the Romans did" and took a bath whenever opportunity offered.

We were situated about midway between Bethune and St. Pol, two towns that did offer some little relaxation from the affairs of war, both physically and mentally. These towns had not been subjected to bombardment like most of the places near the line, apparently as the enemy had visions of capturing the coalfields intact. The population, consequently, had grown considerably beyond their normal numbers, and the busy shops and pleasant old world air was restful after the somewhat lean and tense atmosphere of the aerodrome and Bruay. The eyes felt rested to gaze on neatly dressed feminine France, and although the language was never fully mastered, we soon gained enough experience to make our requirements known. They were simple anyhow.

Our town only contained one fairly decent Estaminet, the *Cerniclet* and this was always popular, one minor reason being, I suspect, because of Odette, the most attractive daughter of the house. Odette's *Maman*, however, had heard all about soldiers, especially Australians, and some years of war had added to her knowledge. Consequently, many there were who tried to gain a place in Odette's affections, but a good meal — for which they invariably paid, was their only solace. After all, a satisfying meal in those days had many things to commend it.

Christmas was almost on us when we arrived and Matt, our Flight Commander, suggested that Christmas dinner should be taken in the village, but we were unfortunate in our lack of knowledge of what could be produced. Other troops, aware of the limitations of the entertainment offering, had booked up the dining accommodation of the *Cerniclet* and we were forced to dine in a very second rate hotel on tinned foods and the everlasting eggs and chips, aided down with copious draughts of the wine of the country. As a meal this combination had little to commend it, but it helped to bring us together as a team, especially during the small hours of the morning. We had to exercise all our powers of diplomacy and tact on the way home to avoid a brawl with a crowd of Tommies who were asleep on the floor of a lorry on their way back from the line. All that occurred was that we heaved about a dozen snowballs into the back of the lorry as it passed, not knowing that some troops were trying to get in a bit of shut-eye on the floor. One bright laddie named Bill, whose history and habits were written in the shape of his mutilated ears, wanted blood — "Orfficers or no bleedin' orfficers" and he was ably abetted by his pals. Apparently Bill was a bit of a dab with his dibs, and a trifle livery of disposition. What was more, he was perhaps entitled to feel a little put out at being rudely awakened by about a pound of snow down his back. As they adopted a bellicose attitude, we all without words instinctively agreed that it would be much better to reserve ourselves for the enemy than to be knocked about by our own side. Tab Pflaum, one of our party, adjusted the matter by the timely production of the great panacea for all ills in war — a bottle of rum — and suggested that Bill try a drop to allay the ill effects of the snow. This worked the oracle and everyone parted happily after the contents of the bottle had disappeared.

Fortunately we found one oasis in the social barrenness of the town. Just across the road from the aerodrome was a small shop, half *estaminet*, half private house, run by a Madame Brunêt, a plump elderly woman. Madame was a most pleasant and kindly personality who could never do too much for us. She adopted us from the start and never tired of exclaiming upon the remarkable fact, to her, of Australians coming all that distance to help France fight their natural enemy. We could point out as much as we liked that it was our war as much as it was that of France, but it never altered her astonishment nor appreciation. She made her kitchen our club and although we had the run of the house, and a small dining room was always available for meals, we preferred the homeliness of the *cuisine*. The old lady undertook the mending of our socks and clothing from the outset and supervised the laundering of the linen of those of us nearer to her heart, and if any of us caught any of the milder ailments, she would be on the job immediately with her simple remedies. Her husband had been a mine superintendent in one of the Lens coalmines, but when the Germans had come through the district in 1914, they had taken him prisoner with many other mine officials and he was doing something in the coal business as a prisoner of war. Madame hated the Germans with the fervidness that only the French can sustain, and anything in her power that she could do to damage him in any way, she would do. That we were fighting him in the air and bombing his troops, was sufficient for her to take us to her ample bosom.

Her assistant, Leonie, was a strong cheerful girl of peasant stock, with a remarkable ability to put people in their places. She knew a fair amount of English and she shared with Madame her hatreds, and between them we were well looked after. Practically every night we partook of eggs, chip potatoes and coffee, in the little kitchen before we turned in, and there was always a hefty lashing of rum for anyone who had had a successful day. We all became particularly fond of Leonie. Her cheerfulness was always infectious, and a very ready and sharp wit made one keep on one's intellectual toes, if sharp repartee were indulged in. She wasn't by any means beautiful, but one was always attracted by her spotless cleanliness. If you desired a first class row, all you had to do was to crumple her stiffly starched apron, or spill something on the plain board kitchen table. Then sparks would fly and a knowledge of French was not necessary to understand the tenor of her remarks.

George Malley and I always were the favoured ones of the household and "Zheorge" and "Ah'ree" (which was the nearest they could get to Harry) could do little that was wrong. There were two other members of the family, Leonie's mother, who lived in a tiny cottage at the back of Madame's, and Alphonse — Madame's seventeen year old son. After we had been at Bruay some little time, Alphonse found it expedient, *la consequence d'une inconsequence* to get married. He was too young to be called to the colors, but in other directions he had assumed a man's responsibilities. Madame was a bit upset, of course, but with quite a sporting gesture threw a party — and it was some party, notwithstanding the limitations of space. The new wife was absolutely uninspiring, even as a bride, and Alphonse's homely mug was quite a restful sight alongside Ninette's. Lord knows where all the food and wine came from, but there was more food than the guests could accomodate, and great variety. I proposed the health of the bride and bridegroom and I am glad they could not understand all I said — but Leonie pretty well did. A local with much gesture, mostly of an intimate nature, did the parents of the bride and groom.

Then George responded on behalf of the parents, and all speeches were received with acclamation. The party continued until the bride passed out with her head on Alphonse's shoulder, while poor old Alphonse himself was asleep with his face in his plate. Frank Felstead, our equipment officer, had stuck a cigar in his ear. The *pièce de resistance* however, was the putting of the bride and bridegroom to bed as they could not do so themselves. They were both absolutely limp and of no mean weight, and they had to be carted out to Leonie's mother's cottage at the back, where they were to occupy a room. It took about six of us to get Ninette out and it was surprising how many different parts of an unconscious woman can bend, especially when she happens to be rather full bodied. You can get their feet and shoulders up, and the stern hits the ground, then you can get that clear and the backs of the knees seem to bend in the wrong direction. There is nowhere you can get a firm purchase and they tend to flow in all directions.

We finally got them both on to the bed, but they could not be left like that on their wedding night, so we got Madame to supervise undressing them. We put the girl's foundation garments on Alphonse, tied her stockings round his neck and touched up his features with a bit of burnt cork and then painted an imitation of the R.F.C. cockade of red and white and blue circles on his tummy with black and tan boot polish. Our modesty forbade us playing about with the body of the lady, other than giving her a nice burnt cork moustache, and so we left them.

We called next morning to see how things were, but although Madame and Leonie were almost hysterical with laughter, the newlyweds had a different outlook. Alphonse was alternately crying with the pain in his head and trying to swear in English. Even the administration of brandy and black coffee and a couple of aspirins apiece failed to ameliorate their discomfort, so, after admonishing them on their waywardness, we left them to their own devices. As a wedding ceremony it was a great success, but as a marriage it was a washout.

In the next chapter, Cobby tells of his 'Charlie Chaplin' aluminium cutout figures, one for each side of the fuselage of his Camel. About 40 cm (16 in) high, the starboard one is depicted here. It is now on display at the R.A.A.F. Museum at Point Cook, Victoria.

Serious Work

Camels of No. 4 Squadron A.F.C. at Bruay on 26th March 1918. The machines had just been prepared for further operations against the advancing Germans. In the background were Bristol Fighters of a Royal Flying Corps unit operating from the same airfield.

CHAPTER 3

"A" Flight had as pilots, in addition to G.C. Matthews the Flight Commander, Tab Pflaum, Bill Brasch, Burton, Willmott and myself. Matt was a Scot, and had been the first mate of a merchant ship, having come to the Air Force from the Light Horse. Tab was a farmer and stock and station agent in Australia. Bill Brasch was involved with cars. Burton was an engineer, newly graduated from Sydney University. Willmot was also in the car business, and of course I had been a bank clerk. They were a very good team of chaps, possessing all the physical characteristics one looks for in one's friends and had already in many ways proved themselves the sort of fellows who would stick together in any sort of situation. With the exception of Burton, we were kindred spirits and all liked to kick up our heels now and again, preferring to do it in company. Roy Burton kept to himself more than the rest and gave a lot of time to playing patience. He did not keep going very long, but was invalided back to England, but it is remarkable to realize that now, twenty three years after the war, we are all alive and kicking. We all saw a good deal of the war together, notwithstanding that at that time we were collectively about the most ignorant batch of pilots, as far as knowledge of air fighting went, that you could find in France.

Practice flying, gunnery from the air, learning the lie of the land and the geography of the front line, were the tasks of the first week or so, and during this period the greatest surprise to us all was, I think, the entire absence of anything pronounced in the matter of a front line. Imagination had always led me to visualise miles of bursting shells, masses of smoke, and the line itself defined with some boldness by a continuous series of earth works, stretching from the coast of Belgium to wherever the war ceased to the south. Actually it was very much to the contrary. A most tranquil state of affairs existed in the vicinity of the trenches during the daytime. Looking down from anything above two thousand feet, all one could see was a long irregular smudge gradually extending out to cultivation. The smudge itself was only ribbonlike, but the condition that left the eye groping unconsciously to account for a feeling of strangeness was the entire absence of movement for miles on either side of this "ribbon." An absolute stillness pervaded the scene. Nothing impressed itself upon us more that this at the time, but it was not always to be so. From the aerodrome at night, one could of course see now and again a star shell, or parachute flare, hover in the black vault of the sky over the line, and sometimes the rumble of guns reached us in our quarters, but there was none of the pomp and panoply of war, no desperate sorties, as far as we knew, to regain or capture ground. The newspapers, a day or so old, informed us that minor battles were gallantly decided on portions of the front that we were endeavouring to discover, yet nothing but peacefulness was apparent to us. But we were fifteen odd miles from the war when at home, and the fellows in the trenches may have an entirely different story to tell.

Our flight commanders were hard taskmasters during this period. The continual drumming into our ears of the stern necessity for close formation work, and the need to be excellent shots, made us willing workers and we gave all possible time in the endeavor to familiarise ourselves with these most important phases of a "scout's" job. A moderate estimate of the amount of ammunition used per pilot per day, on these practices, would be about two thousand rounds, all fired at ground targets at the end of the aerodrome. In these first couple of weeks, we all put in about four times as much flying as we had during the whole of the previous year.

Strangely enough with all our forebodings about our ineptitude to meet the enemy on an equal footing, our first casualties were not the result of enemy action. Hughes-Chamberlain, the O.C. of "C" Flight, was leading a formation of five along our side of the line having a "look-see" and showing them the sector that we were going to be in some measure responsible for, when he put the formation into a steep dive. On flattening out, Anderson and Curtis collided and their machines fell on another member of the flight, a chap named Cash. All three were killed and the atmosphere was gloomy and depressing for a day or so. They had been such excellent fellows and to go out in such a simple manner, seemed a waste of good fighting material, quite apart from the personal side. Thereafter steep dives were watched very closely, and we all endeavored to make sure where the other fellows were before pulling out.

Then the serious work started. Our first job came along on the ninth of January, when we had the rather prosaic task of escorting a photographic machine into Hunland and back again. Matt led the flight over on this escort in diamond formation, with myself as last man of the diamond. I spent the whole of the time looking behind me or into the sun, with only an infrequent glance ahead to make sure I was still in position. Nothing occurred to worry us, however, and we landed after about an hour and a half's trip. We naturally felt bucked at having done the first patrol over the line, small as it was. Matthews did say that we stuck so closely to him that he was frightened at times to turn his machine for fear of running some of us down, but that did not matter — we were home. Bill Brasch also had a few caustic remarks to pass about the way I was skidding sideways at him now and then, to his extreme discomfort, but he blamed my mascot for that. Before I left England I had cut out two figures of Charlie Chaplin in aluminium, painted them according to the latest screen version of his get-up, and screwed one on to each side of the fuselage. I had also "borrowed" two notices from a train in England, one "CAUTION — IT IS DANGEROUS TO LEAN OUT OF THE WINDOWS" and the other "PLEASE DO NOT SPIT". These graced the positions above and below Charlie respectively. Bill reckoned that you couldn't help skidding with those ruddy things on, so we let it go at that. But it was realized that the skidding habit had to be stopped, as it was most dangerous as well as demoralizing in formation, so we gave some extra time to disposing of it, and it was gradually eliminated.

Soon after this, on the 13th, we lost our first man over Hunland. Matt was again leading and Burton, Willmott and I comprised the remainder of the flight. I can remember Willmott industriously painting his college badge on his machine before he came to France, with its motto "Resurgam" which I understood meant "I will arise again". On this occasion it did not live up to its bold assertion, for when called upon to rise, it was the one thing it would not do. We had just about finished an offensive patrol, and were still over Hunland on our way back to the lines, when three enemy Albatross Scouts dropped down from above and behind us and proceeded to take potshots at the rear machine, which happened to be mine. It was a most unnerving experience as the leading machines of my patrol did not appear to notice them and I certainly did not know what to do. The enemy did not come close, and had I turned to attack them I would have been on my own against three, odds which I very definitely was not prepared to give in those days. The decision, however, was more or less taken out of my hands. Willmott's engine had several times given trouble during this trip, and he had lost height before it picked up again, and it now proceeded to given an example of perversity, by throwing out puffs of smoke, which suggested valve trouble. He immediately began to lag, and I was in the uncomfortable position of having to stay with him while the Bosches kept up the potshot business. He was losing height, as well as dropping further and further back, so I zigzagged over him as he went down. The Bosches still hung back, but kept up firing bursts at both of us, when for a moment Willmott's engine picked up again and he started to climb. Our leader and Burton were miles away by this time and had apparently not missed us, having left the job of sticking to them in our hands, a requirement that had been so often drummed into our ears. I was doubtful of the part I was playing in the little drama. I was not sure whether I should have stayed with the formation, or boldly attacked the enemy. Meanwhile I was skidding my machine about and above Willmott, with the intention of not over-running him, but mainly to avoid being shot myself, and wondering principally what was the safest thing to do, when his engine suddenly threw out a cloud of black smoke and his machine went into a spin. The Bosche did not attempt to go after him so I did a "split-S" for the line. Apparently Burton had seen the episode and had managed to convey to Matt that there was something wrong, for I met them coming back. It was too late to do anything then, so we went home. Burton told me that he had seen our predicament, but did not know whether to come to our assistance, or to stick to the leader as he had so often been told. I would have felt a lot more comfortable if he had turned back. We might have saved both pilot and machine if he had, and would probably have got one or two Albatrosses as well, as they did not seem to be anything but very ordinary performers; otherwise they would have put up a more dashing display. We heard later that Willmott was a prisoner and uninjured. The Commanding Officer wanted to know why I had not attacked them, but I had to admit quite frankly that I did not know what to do. As things turned out, I think that I did the right thing. Willmott was alive and in particular, so was I, but I am of the opinion that Willmott might have had better luck if he had put his motto on upside down.

My knowledge was broadening. Apart from the lesson that had been drilled into me about sticking to the leader, the application of which was now a little ambiguous, plus the importance of accurate shooting, I had learned that all German pilots were not the out and out top line pilots that I had been led to believe, and that there were at least three of them as timid as myself. Perhaps there were more and naturally my feelings were gradually reflected in my bearing in the air from then on. Whilst not feeling exactly heroic, I certainly felt that the cold sensation down the back that had hitherto accompanied me on my trips to the line was being dissipated and if I could only be fortunate enough to meet innoncents like myself for a while, I might yet learn sufficient to give a good account of myself in a proper scrap.

Bad weather for a few days gave us further opportunity to practise gunnery and low flying in formation. We developed a keenness in mothering our guns, sights and synchronising gear. The latter was the apparatus that was designed to permit the bullets from our guns to pass between the blades of the propeller without hitting them and nearly always did. It was a delicate piece of mechanism, however, and was easily thrown out of adjustment. After every flight we used to check these things over, first to assure ourselves that the guns were aiming at the same spot as the sights indicated, and then to see that the gun gear fired the bullet the right distance behind the trail-edge of the blades. If this was out of order, you had a good chance of shooting the prop. off in the air, with probable dire results. The proof of the test was the size of the burst on a target at the "stop butt". Another matter we were very careful of was not to impose undue strain upon our buses by idle stunting during these practice flights. Seldom indeed did we put them over in a loop even, but reserved their strength for those occasions over the line when they might be called upon to stand up to all sorts of stresses in combat, or in pressing home an advantage. Further, there was the chance that a stray bullet may have found its way into some vital part of the structure without being noticed, although with the rigid examination to which they were subjected, this was unlikely. But we were all a bit "windy" about the risk of a break-up in the air.

We were also gradually developing a pride in our formation flying, its excellence partially induced by the biting remarks of our pals if a mistake was made, but mainly by reason of the fact that we were learning in the strictest of all schools — that of necessity. The casual remarks of pilots of other squadrons "that you ruddy Australians can formate" (a word coined in the Flying Corps to embrace all types of formation flying) made pleasant hearing. In those days, units of quite considerable experience were not particularly exacting in their air drill, but from the first it was insisted upon in No. 4. Perhaps it was so by reason of the fact that we were all so new to the game, and close herded instinctively for safety, that we regarded it with so much importance. Whatever the underlying cause, it imbued us with a sense of added security to be able to take off and land again with wingtip tucked in behind wingtip.

Our evenings, when not on late flying, were full of the gossip of outstanding fighters of both sides. The manner in which Ball, Bishop, McCudden and Mannock (who was then receiving notice as a destroyer of enemy machines) staged their attacks, their favorite positions for opening fire, the tactics they employed if surprised, and so on, were never ending subjects of argument and sometimes acrimonious discussion. Many were the theories as to what we would do under different circumstances, and many were the viewpoints for an against them all. The notorious Baron von Richthofen was our principal bogey. He seemed to be unassailable. His victims were already numbered in the sixties, and there seemed to be no way in which to stop him. Evidently he was a dead shot, an excellent pilot and a man of superb courage and still in his early twenties. The combination seemed too much for us lesser mortals. It made us even more determined not to lag behind when on patrol, but other than that we could only leave it to chance, or whatever might be in charge of our destinies at the moment. We had become fatalists and what was going to happen would happen, but we had no intention of throwing in our hands and we would take things as they came. We had learned something more, that the favorite direction of attack by all leading scrappers was from the sun. This gave the advantage of surprise and height to the attacker, and hence-forth we included the luminary orb as one of the things to bear in mind. It was Richthofen's favorite method. He would wait up above a dogfight until one of the enemy pilots became separated from his comrades, and he would dive onto the unfortunate straggler. His celebrated boast that everything below him in the air was as good as finished, was well known to us. He always patrolled in company with a number of other crack German pilots and they all had crimson colored machines and were known as the "Circus".

We recommenced patrols after a few days of dud weather but nothing of an exciting nature occurred other than the pursuit of a two-seater on one occasion, and the shooting up of enemy trenches just for fun from the safe height of about three thousand feet. Be assured that we nearly twisted our heads off watching the sun and in keeping an eye on our tails. We were in no danger of forgetting our lessons.

Then the squadron got its first Hun. Arthur O'Hara Wood attacked an enemy two-seater and crashed it near La Bassee after a short fight. The squadron had drawn blood, and we were most anxious to know all about how it was done, but it appeared that there was nothing to tell. O'Hara Wood just remarked that after a couple of bursts the machine had gone into a spin and hit the deck. We had to guess the rest. We had more or less taken it for granted that he would soon be prominent in the picture, as he was of the type that always looked for trouble in the air, and by temperament was most suited to meet it. He was an athletic chap and very keen on keeping fit, and used to insist on the rest of us getting in a fair amount of exercise. He was the Victorian tennis champion at home, but in France the nearest he could get to his beloved game of tennis, was badminton, but he managed to speed this up by clipping down the feathers of the shuttlecock and making the court about half as big again, and then playing as though he was on the centre court at Wimbledon. Another of his pastimes was boxing, in the mess after dinner at night, but he did not get many takers at this. His regular opponent was Hughes-Chamberlain, and from a gentle beginning a friendly bout would develop into a full-blooded sort of affair with the usual amount of gore about. We others, looking on from armchairs, always enjoyed these shows.

We carried on with offensive patrols, or "o-pips" as they were called, for some days without any particular action. Many times enemy formations were sighted, but they were either too far off to force into battle, or perhaps they were not in a fighting mood. Anyhow nothing eventuated. On all these patrols we finished by firing several hundreds of rounds into the enemy lines or billets, not with any hope of doing particular damage, but just to let him know there was still a war on. Our friends in our trenches frequently told us what a rotten feeling it gave the troops to be fired on even back "in rest", so as a matter of habit we gave the Bosche some just for luck. I often wondered what kind of trouble we gave him and gathered from reports that sometimes filtered through to us, that it was most annoying if nothing else. Now and again we would catch something moving on the roads and then we would give them a little hurry-up. The results on these occasions were probably only slight, but they tended to build up a spirit of aggressiveness that was a most valuable asset.

On February 3rd, Arthur O'Hara Wood was the cause of me getting into my first scrap. I have an idea that he thought it was about time "A" Flight got into some serious business, and one day when Matt was off color, he said that he would take some of us over the line for a bit of a gallop, with the object of catching some unsuspecting Hun looking the other way. He nominated Tab and I to accompany him, and gave us some very strict instructions about sticking to him and doing as he did. We were a bit uneasy about deliberately going out to look for fight for we knew what an aggressive kind of cove old O'Hara was, but we could not do anything but look the part, even if not consumed with eagerness.

"If we get home without being shot up the whole way", said Tab, "I'm a Dutchman, and I'll buy the boys a drink." I myself would have been quite content to get back in one piece and forego the drink.

We left Bruay at somewhere near twenty-five past nine on this stunt, on the morning of February the third, and climbed toward La Bassee, which was on the line about fifteen miles away, and by the time we arrived there, we were about twelve thousand feet up. Wood was in front and still climbing with Tab and I behind and to either side of him. We kept on climbing and going further east, until we two had difficulty in maintaining the same angle of climb. I might mention that it is much harder for those following to keep position and climb, than it is for the leader just to go on climbing, especially if you are climbing flat out, the tendency being to try to climb too steeply with a consequent slowing up of the rate of climb due to partially stalling. The result is that every now and then the machine goes into a spin and the struggle to regain lost height starts all over again. O'Hara was manoeuvring round and round all the time he was climbing and was frequently out of sight, being blanketted by our top wings, and this coupled with our several spins, in which we lost a couple of hundred feet each time, had made us both a bit "bloody minded." Tab had just gone into another spin and was about two hundred feet below me, when suddenly Wood kicked his machine over into a dive to the left. The fact that Tab and I were so much below the leader now became an advantage, as we were able to get into position without much loss of distance and we went "lickety-split" after him. I had no idea what the business was all about, but I clung to the tail of the machine in front. It seemed funny to me to be put to so much effort and time into getting to such a height (we were somewhere near seventeen thousand feet when we started to dive) to throw it all away again in this dive in which we were indulging. All in a moment I saw the reason for it. Right ahead of O'Hara Wood was an enemy two-seater and he was diving right at it. A little further away and scooting east for all it was worth was another one, so I moved in its direction. It had no chance of getting away without my having a shot at it owing to the tremendous speed that we had developed during our long dive. It was by far the fastest that I had ever dived a machine before.

The enemy, which I now discovered to be a D.F.W., a type usually employed on photographic work, was about 2,000 feet below me and to the east a little when I first saw it, but it was only a matter of split seconds before I was almost on top of it and approaching from its left hand side. Then suddenly there appeared to be too many things to do at once. I had to ease off my engine to steady the vibration

"A" Flight's sergeant armorer poses in front of "Charlie Chaplin" whilst the aircraft's gear receives attention. The white vertical identification bar replaced the former white boomerang during April or May of 1918.

'Tabby' Pflaum, one of Cobby's closest friends. A wonderful character, he was a solid performer and a good team man. This photograph shows him as a captain. The picture was obviously taken toward the close of hostilities.

down, otherwise I might shoot my propeller off owing to the way it was shaking, and my bullets would have gone all over the place; then I had to sight my guns and allow for the angle of attack and the speed of the enemy machine. I found that I had started a bit too late to settle my sights on the right place, and had to put on right rudder to bring my guns past the pilot. I pressed the triggers when only about twenty feet away and just managed to get one burst in before pulling out of the dive and skimming across with my wheels, a few inches above the wings of the enemy. The German observer had been firing at me all the time I had been coming down on him, but as I pulled up into a zoom I saw him slump down in his cockpit. As I had half expected, I had not aimed off sufficiently to get the pilot, but knowing that I was safe from the rear gun, I half-rolled and came down again, but this time dead behind and just below the tail of my opponent. I fired four or five bursts at close range from this position, but nothing happened and the enemy machine just tootled off on its way home. I then thought it time to look around and see where my comrades were, but not a sign of them could I see and suddenly I felt very much alone. I could not recognise where I was, but from the position of the sun I knew which was west, so I pushed off towards where I gathered the line was. After some minutes I saw another Camel down below me, flying roughly south, so I went after it, imagining that it would be O'Hara Wood, but even using all my engine I was unable to come up with him and get into a position where I could see his identification marks. Presently he circled as though to land and I shot across in front of him and to my extreme surprise it was my old pal Tab. Then — whizz! — past my nose shot another Camel and this time it was Wood. Tab and I formed up behind him in our appropriate positions and were led home like a couple of lost lambs, making good signs as to wetting the scrap in the approved manner when we got home, and describing in mimicry what we had been doing. Unfortunately O'Hara had an idea that we had left him in the lurch, and catching sight of some of our fooling, thought that we were having a joke on him. As soon as we got out of our machines, he stormed over and wanted to know "What the . . . . . hell do you mean by funking the fight and leaving me on my own? You two goats can push back to England and get a bit more training, you are no damned use out here." We were astounded. Tab and I both argued the point about leaving him to carry on by himself and both said our pieces. Tab apparently had had a scrap with another machine and said that it had spun down out of control, but the rest of us had not seen anything of it. I for one was too busy to have had time to look out for other people's scraps, and besides was under the impression that both Tab and O'Hara Wood were looking after the first two-seater, and if any looking after was to be done, I was the one to receive the benefit. I could see that Wood was more or less sceptical of all that we had told him, but he was too much of a sportsman to condemn us out of hand. "Perhaps some of the observation posts on the line might have seen something of what happened", he remarked, "I'll get in touch with the line and see what's what; the visibility was pretty good and we were only as far over as Douai."

Tab and I repaired to the Mess and had a couple of spots in honor of our safe return, ignominious though it appeared.

"I am damned certain the machine I had a go at went down in a spin", Tab insisted. "I admit that I did not wait to see if it crashed, and the pilot may only have done it to get away, but I'm willing to bet a pound that half the line saw it." "You're lucky if they did " was my reply. "In the first place my bloke went off unhurt as far as I could judge, except that the observer was definitely shot up. I am not sure where it happened as I lost touch with where we were, other than we must have been somewhere about twenty miles over Hunland from La Bassée, when I left."

I was upset with the reception that had greeted my explanation, and as far as I could see there was no means of proving the truth of it. Tab would not for a moment admit that one should feel concerned about it. His contention was that as far as he was concerned, he had done as good a job of work as he could and if they did not believe it they could do the other thing. After a few drinks with the chaps, I went down to the *Cerniclet* for lunch. I preferred my own company at the moment.

When I came back an hour or two afterwards, there was a different atmosphere in the mess. Tab and the remainder of the flight were in the middle of a "binge" and greeted me with a chorus of yells. In my absence word had come in from the front line confirming our statements in every particular, except that in my case I was not as far over as I had thought. What was more satisfactory still was the fact that *we had all got our machines*. Tab had crashed his just across the line from La Bassée, O'Hara Wood had shot his down near Brebieres and mine had burst into flames just near Douai when still about a thousand feet up, apparently having done so soon after I had turned away from it.

The reaction was almost as bad as the depressed feeling of the morning. We all had bad heads next day, but they were worth it to have merely regained our standing with the rest of the squadron. It was the first "notch" to the Flight and we had our tails well up in consequence. To get three Huns in our first scrap was much more than we had ever dreamed of, and the thought left an abiding feeling of well being with us.

In the centre, Rittmeister Manfred Freiherr von Richthofen, the famous Red Baron. On his left were Lieutenant-Colonel Erich Lowenhardt, then Lothar von Richthofen, the Baron's brother. (The others are unidentified.)

Richthofen's Circus

CHAPTER 4

The squadron now settled down to combined bombing and offensive patrols. Every machine now carried four twenty-five pound bombs, designed for action against enemy aircraft, as the High Command, or whoever it was that controlled such things, ordered that we should extend our duties to that of bombing any suitable target that might present itself whilst on patrol. Needless to say, we usually got rid of this added weight in the direction of the first suitable object that came along. Sometimes we were caught by the enemy with them still on and had to discard them in a hurry, taking good care that we were over the enemy's lines. They were too much of a handicap for our little machines to lug around in a fight. Sometimes our orders were that we should definitely bomb some part of the enemy line or back area, where it was known that activity was taking place. On these occasions we were usually "Archied" the whole time we were in range, a proceeding that, to say the least of it, was uncomfortable but it has been more tersely expressed than that. Anti-aircraft fire, or "Archie" as it was more cryptically known amongst pilots, was always the bugbear of new pilots. It was responsible for the cracking up of many of the chaps who were invalided back to England. It was not that it claimed very many victims, as the result of hits, but it seemed to have a kind of hoodoo on beginners. As a matter of fact, after a while we came to regard it as something that annoyed us but had to be put up with. It was always with us. It was not treated with contempt, but once on patrol, by the careful watching of the time we stayed at any one height or flew in any one direction, we were able to render it fairly innocuous. It was a source of interest to some of the mathematically inclined, to work out what it cost the enemy in shells after every patrol. A fairly low estimate of the cost of putting each shell into the sky would be about six or seven pounds, and when you consider that perhaps a couple of hundred shells might well be fired at a patrol in two or three hours, you can see that even though not destroying the enemy in the air, a lot was being done to make the war very expensive for him. At least some of our chaps used to regard it in this light. Another anti-aircraft weapon was a contrivance that fired a series of things that resembled a lot of red-hot eggs strung together. These were commonly known as "flaming onions" and I do not know whether they ever accounted for any machines or not. Pilots were now and then reported as having been shot down by them, but personally I think they were fairly harmless, and that their value lay in the extremely demoralizing effect of meeting fire in the air.

"B" Flight had some luck about this time in shooting down two of a patrol of six Albatross Scouts, without losing anyone, but in other dogfights the squadron lost three men from "B" and "C" Flights, with two taken prisoner and one shot. We were now doing a lot of flying and it was beginning to show its effect on some of the pilots. We were concerned that one or two were showing signs of strain, and the thought that they were probably committed to the job of flying until the war came to a conclusion at some indefinite time ahead, was having a bad effect. One or two were developing a nervousness that was a sure sign of an impending breakdown. Already they had indicated that they could not stand up to the job much longer, and the only thing to do was to send them back to England for a period. One of the most difficult decisions for a commander to make is that of deciding whether a pilot is getting nerves or merely cold feet. In either case he is a demoralizing influence and should be removed, but the method of doing so depends so greatly on the appreciation of his qualities. Usually the flight commander can judge from constant association what is the trouble, and can help the man concerned by putting him on to easier patrols for a while or by nominating him to act as his own fighting partner so as to give him the advantage of moral support. Sometimes a talk will fix the business, but all one's tact has to be employed to prevent the unfortunate fellow from feeling that his condition is apparent to others, as it is so easy to make him go off the rails through the careless use of words. One particular case that I have in mind, was that of a chap in one of the other flights who had been doing excellent work, when something happened that made him so jumpy that he could not force himself across the line. He was chipped in front of the whole Mess one night at dinner, and broke down. Next day he was a physical and mental wreck, but said that he was going over the lines to finish it, and off he went, a peculiar anomaly of a man who temporarily was definitely too frightened to cross the lines with a number of stout chaps, because of the danger of being shot down, yet was going over by himself with the definite intention of being killed. The conclusion was entirely different from what we anticipated. Seemingly he had gone well over Hunland looking for a fight and after some time had met a number of enemy scouts which immediately attacked him. His own words were, "I felt almost helpless with a kind of terror and could not handle my machine at all. After a while I noticed a burst of bullet holes in my left hand wing, and the sight of this woke me up and I started to pull myself together. I found that I could outfly the Huns so commenced to force the fight against them a bit. I was unable to get a decent shot at them, and after a few moments they gave it up and cleared off. I chased them but they were too fast for me, and they got away, so I had a look around for any more, but seeing none, I returned home." As a matter of fact we all had half an idea that he was going to land on the Hun side of the line and give himself up. The C.O. suggested that he might care to go on leave for a while, but he was most indignant and insisted on going on with the job. He was as right as rain afterwards, but admitted to me that his trouble was that he only felt nervous when being led by a certain flight commander.

I think I only met one coward during the whole of my flying experience at the War and he was to be pitied. I knew only too well my own earlier feelings, and this chap would have surmounted his fears if he could only have been made to get into a decent scrap and get shot at a few times. I was always confident that this type of chap would soon get his "paddy" up if some cheerful Hun had a few cracks at him, and temper is frequently mistaken for bravery and a raging temper isn't a bad substitute for heroism. If you are properly mad you haven't room in your thoughts for fear.

We met some trouble near the end of the month but not at the hands of the enemy. Following a scrap in which thirty or forty machines of different types on either side were involved without, strangely enough, any casualties being inflicted, we had scattered in our various formations and headed back for the lines. We were well up somewhere about 17,000 feet and could see a dense bank of fog or cloud moving up from the coast, but our height prevented us from judging whether it was on the ground or whether it was a cloud. After a little while we decided to go down and have a look. We were surprised to find it was a very dense fog. In a few minutes we were out of sight of everything except this wet grey blanket and in no time our machines became separated. Every now and then someone would whizz past in the murk. From ground level nothing was visible twenty yards ahead, and being afraid of colliding with houses and trees, if not another machine, I climbed up and out of it, and after waiting until my compass had settled down, made off for the direction of Bruay.

Bruay, too, was under fog, so I continued on hoping to be able to see through presently, but for another half hour's flying it showed no lessening in density and so I attempted to get through and find somewhere to put the old 'bus down. My petrol was getting low and if I didn't do so pretty soon, I would be forced to land. I reasoned that I would be about thirty miles South West of Bruay, somewhere between St. Pol and Abbeville, and gliding down very slowly, I watched my altimeter creeping towards zero, when all of a sudden, as the instrument showed that I was still about two hundred feet up, there was a crash at my right wingtip and everything faded out.

I woke up completely boxed mentally and physically. I was on the back of my head in snow, gazing up alongside the fuselage at the fog. All except my head was inside the machine and I was completely upside down and held down by the weight of the machine. The cutaway side of the cockpit just accomodated my neck and allowed me enough room to breathe, but the pressure, though not severe, was just enough to prevent even my Adam's apple from functioning, and as I felt petrol running down my legs and thence down the back of my neck and had visions of fire, the aforesaid Adam's apple tried to do a little overtime. There were no wings on my side and my range of vision was limited by my position and still further by the fog and snow, and a large amount of hay that seemed to be clinging to every excrescence on the machine. I was badly bruised, but seemed all right otherwise. My lungs, however, were intact, but it took about half an hour's yelling *Ici!*, *Au Secour!*, *Blessé!*, *Aeroplane frappez!*, and all the odd bits of French I knew, to indicate I wanted help before anyone arrived. Then a very startled old man of the peasant type appeared through the fog, and if I was frightened of fire this old chap was very much more frightened of me and my machine. I never did know much French, but could always make myself understood as far as the everyday requirements of life went, but I never realized how true was the oft repeated jest that certain races would be dumb if you tied their hands behind their backs. My arms were inside the machine and twisted up in the works a bit, and only my bloodstained face was protruding. I had split my lips about a bit. My repeated "*Ici, s'il vous plait*", "*Au secour merci*", which was intended for "Come here and help me please", had not the slightest effect on him. My attempts to get him to break up the machine by repeating "*Allez — frappez machine, encour bombard machine avec votre pied.*" which was my best effort to get him to "Go on, break up the machine, kick it to bits with your foot." alarmed him even more. I then tried to tell him to fetch a gendarme if there was one about, or to bring along any soldiers, but could not think what "bring" was in French. My best was "*Donnez moi beaucoup gendarmes*" "*Donnex moi beaucoup soldats*". He brightened when I mentioned gendarme, and after I repeated "*Allez*", he shuffled off. About three-quarters of an hour later he reappeared with the local representative of law and order, in the form of a corpulent and red-faced Gilbert-and-

Sullivan type of gendarme, followed by an excited string of villagers.

After a voluble discourse amongst the lot of them, accompanied by much expression of face and waving of hands, the gendarme addressed a long speech at me, which was either an address of welcome to their humble village, or an instruction to do nothing until he heard from the *Sureté Generale*, or whatever his department was. However, I knew nothing whatever of what he said. My ponderous "*Merci monsieur*", "*Felicitations mon brave gendarme, allez vous avec machine atil pas bon ici*" must have penetrated to the grey matter that had earned him the job of local peace officer. His face beamed immediately and one could see that he suddenly realized that the posture of the machine and myself, was not one adopted to win a bet, and in true Napoleonic style he issued orders like the rattle of a battery of machine guns. Before I realized it, people were heaving themselves against the machine, while others were trying to pull me clear by the head, but although I lacked the knowledge to tell them in their own language to desist, average Australian, plus emphasis, stayed their eagerness in that direction. Presently they had tipped the machine over enough to disentangle me from the wreckage. I was surprised to find that I was practically unhurt. Except for my lacerated lips and a torn bump on my forehead where I had knocked myself out against the gun butts and a lump at the base of my skull, I was practically intact, very stiff and cold certainly, but able to walk. I went back a bit to examine the mode of my arrival in the upside-down position and found that I had struck the branches of a tall tree about forty yards back, which had pulled me down. I had then, of all things, hit the top of the only haystack for about a couple of miles, and then somersaulted the remaining distance. A pair of crumpled wings at the foot of the tree were mute reminders of the closeness of things. It took me about a day to get back to the squadron, and found that the other chaps on patrol had had forced landings all over the countryside, some of them in other squadrons fatally, and several of them smashing themselves up. Bill Brasch was the only one of our squadron to get back the same day.

Our strenuous period of work now brought us a small measure of reward. We were ordered to proceed by flights, pilots only, to Berck-sur-Mer for a couple of weeks rest and ground gunnery. I have never discovered whether this was a kindly act of grace by the "high-ups", or whether they reckoned us such a rotten lot of shots that they had better do something about it, or whether it was a kind of preliminary fattening up preparatory to the final kill. Whatever it was, it suited us down to the ground. We had heard all about Berck from our opposite numbers in No. 2, how it was a beautiful seaside resort much patronised by invalids. We had been told of its proximity to Berck Plage, Paris Plage, and other delectable spots hitherto unspoilt by the ramifications of the brutal and licentious soldiery, and also of the many distractions the district offered to tired and war-torn pilots.

This gunnery school would be a delight to any youth, but other than being amusing and diverting, possessed little value for training for the serious business of war flying and fighting. There were metal rabits on an endless chain that galloped along the top of a row of sandbags, which you were supposed to hit with bullets fired from a Winchester rifle, mounted in a balanced fuselage, which one controlled by the rudder bar and joystick. Sometimes someone hit the rabbits. Then there was a metal silhouette of an aeroplane that zigzagged down a steel tower in a mad scattering scramble, designed to train the eye in allowing for the speed of a machine in flight. The same fuselage contraption was used for firing, but I don't think anyone

Lieutenants J.H. Weingarth, R.G. Smallwood and V.G.M. Sheppard pose for the official photographer on 15th May 1918, whilst at Clairmarais, France. The Camel was trestled for harmonisation of sights and guns. Amongst the upturned earth near Weingarth's right foot is a heap of expended shells and belt linkages. Note the bomb racks for four 25 lb. Cooper bombs.

ever hit the target. Other ranges for miniature rifles were available and were used daily, but in a light vein rather than in a serious way. The king pin of the place for amusement and risk to the onlooker, was an arrangement designed for training in the use of the observer's gun. Being single-seater pilots, we did not use observers, but that did not matter. This gun, a Lewis, was mounted on an ordinary aircraft revolving mounting and could be swung, by considerable effort and loss of skin, in any direction. This apparatus was in turn erected on a shoulder high "cockpit" on a four-wheeled trolley of about two feet wheelbase, and was towed by another small motor (unmanned) at about 20 miles per hour around a continuous and very small railway. The target was fixed to another motor on a similar track adjacent, and travelled in the opposite direction to the "gun". The job was to hit the target, correct stoppages, reload, etc., while your mount careered around the track, bumping and swaying and jerking in all directions and always at the moment you managed to get your sights on the target. At the best one only got a fleeting shot or two in perhaps ten minutes ride, and although one staggered out of the bus with skin off the knuckles and fingers, with bruised and trembling knees, and perhaps a good many bumps elsewhere, it was the safest place on the range. Quite frequently the occupant of the "cockpit" would just press the trigger when the contraption would jerk in another direction and he would spray the landscape with bullets as he swung the gun over to the other side. There was always a sweepstake on the morning's work and the winner invariably invested the result in the wine of the country, and having every afternoon and evening free, life was very pleasant. This particular part of the coast is looked upon as having a climate and or other conditions very favorable to invalids, who seemed to devote the day to being drawn around in four-wheeled "beds" by dogs of remarkable size, mostly Great Danes. It was our intention to take one of these animals back to the aerodrome as a mascot, and during the early part of our stay we gave a considerable amount of our time to studying the layout of the place, and to discovering the home of the most "getatable" dog. We pinpointed two likely animals and resolved that on the last night, one of them should be ours. As the object of our attentions we had selected a massive brute that stood as high as a pony. The selection had been primarily made when its owner, reclining in his dog-drawn vehicle under a sunshade, and who was in the habit of calling down the wrath of heaven on us every time we passed him in the tender, had used some unkind epithets regarding Australians. Apparently the tender made a dust which he objected to, but roads through sand dunes are always dusty. His remarks were varied and contained allusions to "pigs", "name of a dog", and other domestic animals and the opportunity to give him a particular dog, or the loss of it, to talk about and at the same time to bring to fruition our early made plans, seemed too great to let slip.

That night the eight of us dined well at the Hotel Royale in Berch-sur-mer. We sang our swan song and publicly embraced everyone we knew, and a good many we did not, and departed in the early hours to collect the mascot. On the way out, Matt thought the sign outside a maternity hospital would look well in camp. It was about six feet by four, and had the name of the hospital and its particular function, which related to the bringing of children into the world in letters nearly a foot hight, round a red cross. It was collected by the aid of a little application of strength, and for months afterwards remained mounted outside the squadron medical hut. Then some other bright laddie thought that the symbol on top of the statue to the heroes of the Franco-Prussian War of 1870 would look well on the farm at home if we could get it. This monument was formed by symbolic figures mostly in the natural, very much entwined of limb, surmounted by a Cupid-like form brandishing aloft a sword, the whole I presume to indicate that all sections of the community were welded together in love of country and would do a little of the last man and shilling stuff before they gave in. There was a nice circular pond with about three feet of water in it all around the statue and a fountain played prettily over the nude figures, male and female. Cupid was a nice plump little chap in bronze, about thirty inches high and would undoubtedly lend grace to the sliprail post on some remote property back home, or so it seemed to us. He was only about eight feet off the ground, and after we had climbed the spiked iron railings, the ascent promised to be easy as there were many protruding parts in the marble figures that promised an easy foot or hand hold. Matt had some hazy idea that the finer susceptibilities would be offended if he trod upon the marble bosom on the Junoesque figures in front of him, but it was pointed out that we were in France anyway and besides we weren't all going to ascend by way of female figures, but that one or two of us would have to utilise some of the male variety and that they might have just cause for complaint too.

I scaled the height by seizing the whiskers of an Ajax-like form and used as a foothold the only prominent feature of the Godlike figure.

The chubby little chap on top was fixed to the base of the column by lead mortising and by three of us applying our weight and heaving, we managed to work the figure loose, but it was keyed in too securely to be moved. Unfortunately we were not silent workers, and what with the "heave yo-hos" of Matt, the encouragement of those below to get on with the job, and ribald comment upon the statutory, a considerable disturbance was made. Windows started to open and epithets of varied description were hurled at us. Apparently we were brigands of a particularly low order, and thinking discretion should be our watchword, somewhat belatedly we retreated in the direction of what was to be the major operation of the evening — the home of "Pluto", the giant mastiff, or whatever it was.

Our quarry was housed in a yard on the outskirts of Berch on the road out to the range, but the yard was surrounded by an eight foot wall of stone. We assisted each other to reach the top where we sat in a row, making overtures of friendship with the dog. He seemed quite willing to be friends, and stood on his hind legs and licked our boots, while we tried to pat his head. This precipitated things somewhat, for in leaning forward somebody lost his seat and as he was sliding over grabbed the ones on either side, and "plonk" three of us found ourselves in the yard with an animal that was a combination of a playful hippopotamus and one possessing the agility of a puma. It then proceeded to frolic and gambol with us. In about three minutes I had been knocked over half a dozen times, mostly up against the stone wall, had had my face licked by a tongue that nearly took the skin off, and had one sleeve nearly torn out of my tunic. It took just that time to decide that we didn't want the animal, but the playful brute was still full of fun. We had to be pulled out by the chaps remaining on the wall and the human posterior being constructed as it is, their purchase thereon was exceedingly precarious. Nearly every time one of us would be on the way up the wall, little playful would leap up and give him a frolicsome buffet with a couple of paws as big as boxing gloves, backed up by the impetus of about 600 lbs of dogflesh in flight.

We did get out ultimately, but were no longer the spick-and-span young officers we had been. Without discussion we had all decided that the morning should not see us in Berck, so we made back for the range as fast as we could. So ended our gunnery training course and back we went to Bruay and the war.

Whilst a Bruay some thoughtful "higher up" keen on the health of his troops had come to the conclusion that pilots did not take enough congenial exercise and had arranged for the supply of several horses for us to ride. This was quite a sound idea, and we took full advantage of it. However the horsemen of the squadron, particularly the ex-members of the Light Horse, considered that the only way to ride was to take the gee-gees across country over everything, it became a bit tough on those of us who were not experts. When it became apparent that patrols were going out a pilot or two short, owing to sprained shoulders and bruised bodies, the horses disappeared. Exercise again reverted to badminton, or walks round the slag heaps, or fosses as they were called, but these activities could not even remotely be classed as congenial.

We had a few changes during February and a reshuffle of flight commanders. Old Matt went home sick, and Tunbridge from "C" Flight was given command of "A" Flight. This was a disappointment to me. Matt had been ill some of the time we had been in France, and was pretty well knocked up. He was much above the age of the rest of the chaps in the flight, and in addition the sight of one eye was affected, a state of affairs that did not help, but he made up for his physical disability by the exercise of considerable native wit. I think the realization that he might unknowingly lead the flight into dangerous situations, from which he would not be able to extricate them, was largely responsible for his decision to accept an appointment at the Training Wing in England. Even with the best of health and the strongest nerves, it is a worrying business to have the lives of your pals dependent upon your judgment. So back to England we saw him go and he went with our good wishes. He had been a thoughtful leader and a good pal as well. He would be able to keep an eye on the training of the chaps who were learning to fly at the Australian Training Wing and would see that they were given the right stuff before being sent overseas. A new pilot, Norman Trescowthick, reported to "A" Flight early in the month, and he was good value.

Towards the end of the month Burton left the squadron for hospital, so two of our original team (the other being Matt) had now returned to England.

March came in with ominous signs. For some time we had been sighting large numbers of enemy machines, but they had kept well away. Now things started to get "hot" and dogfights on a large scale became the order of the day. The technique of fighting also gradually changed. No longer did we send formations of three or four machines over the line, but strengthened them to eight or ten with an occasional full squadron show. To quicken the interest, the Baron was known to be operating on our front. None of us was particularly anxious to meet him, but several were feeling their "oats" and would not mind doing so if we had a picked bunch of the lads with us. As Bill Brasch was fond of saying, "The bigger they are the heavier they fall."

The daily task became a series of bombing and offensive patrols which included shooting up the enemy trenches, bombing every movement we saw near the lines, with an occasional scrap in which honors were fairly even. New pilots joined the flight and were broken in to their work. Robertson who had joined in January, was now doing regular work and Trescowthick, newly arrived, was attached to patrols and taken as far as the line to get him accustomed to the layout of the country. He was a good pilot and soon took his place in the unit. We now had seven officers in each flight and this allowed one or two to stand down from each patrol, a very acceptable state of affairs as we were doing two or three shows a day when the weather was suitable. Unfavorable conditions for a week or two cut down operations to some extent early in the month, and we were confined to low flying attacks on trenches and low offensive patrols. These low flying jobs were full of incident. The flight would be flying under the clouds at a few hundred feet, shooting up ground targets, and getting in return all the "hate" that the enemy could throw up at us with machine guns, field guns and flaming onions. Now and then we would surprise an enemy machine and have a quick snap at it before it could get into the clouds. Then we would be startled by the "pop-pop-pop-pop" of a gun on our tails, and things would be crowded until we had eluded him or had convinced him that he was up against

The famous Red Baron, Manfred von Richthofen, shakes hands with General von Hoeppner, Commander of the German Air Service at that time.

something as good as himself, when the fight would break off by one or the other leaving for more pleasant surroundings. There was not enough room to get into a proper dogfight and they were usually tip-and-run affairs.

Low flying attacks or "ground strafes" as they were known, were not pleasant tasks. We nearly always came home with machines shot about, and having as we did a strong affection for our buses, a good deal of profanity was directed at the perpetrators of the damage. This had the effect of making the war a personal one as regards the particular part of Hunland responsible for the sacrilege, and every opportunity was taken to remind the enemy of the fact. It was no unusual occurrence to have a machine completely written off on these jobs. The squadron was handicapped by the shortage until such time as the machines could be replaced, and this in addition to losses occasioned by crashes and losses over the line, now and again left us with insufficient aircraft to make up large patrols. We had this remedied by increasing the establishment of machines from eighteen to twenty-four.

Every patrol from now on was meeting large formations of enemy scouts. Two-seaters escorted by fighters were being met continuously, and although there were many brushes with them, and several victories were gained by various members of the squadron, it appeared that the enemy were not prepared at this stage to give battle unless forced to.

Although we did not realize it, they were reserving themselves for a big effort and their increasing activity, plus information received from other sources, had evidently given this impression to the higher command. Courtney leading ten machines had a brush with several two-seaters between Lens and La Bassée, but a large number of enemy scouts high above him warned him to clear out while his luck was in. Later on, however, he had a go at some more and his formation managed to shoot down four without loss to themselves. "C" Flight met with five machines, and an above guard of five of ours met Richthofen's Circus at about sixteen thousand feet above Douai. After a desperate scrap with the odds of sixteen to ten against them, they managed to extricate themselves with the loss of one of their men shot down and two Germans destroyed in flames and one out of control. Gerry Schafer was shot badly in this fight, having some hundred bullets in his machine, some of which smashed into the windscreen directly in front of his face and also wrote off his guns. He himself was unhurt, but he found it necessary to spin his machine down some ten thousand feet to the ground to avoid the attack, and when he realised he had imposed the strain of a spin for so long on a wrecked machine, he went all "goofy" and was not much use for some time. All things considered, we were not having much luck in our flight. A number of indecisive contests with odd two-seaters, and continual shooting up of the trenches, billets and villages in the hands of the enemy on our front, were our daily tasks while we were receiving reports all the time from other squadrons of the large numbers of enemy aircraft about. We anticipated that the period of quietness we were experiencing was to be of short duration, and plodded along getting into the air a couple of times a day and looking for the elusive enemy and not finding him, and then venting our frustration on anything that we could find on the ground.

Then on 19th March we got orders to make a reconnaissance of the sector near ours as far as Bapaume, and to get used to the layout of the country in that direction. We were told that an attack on a large scale was manifestly intended by the German High Command, that its actual location was unknown, but that the indications were that it would commence somewhere about Cambrai or Bapaume. I was detailed to lead two formations of five machines each at the crack of dawn the next morning, 20th of March.

So at 2.30 a.m. ten of us struggled into the mess for an early cup of coffee and a bite to eat. It was bitterly cold and the whole landscape was blotted out by mist. It was far from being a pleasant prospect and coupled with the hour, produced a feeling of deflation amongst us, which did not improve as we hung about waiting for the weather to clear sufficiently for us to take off.

Just before eight the mist had dispersed sufficiently to warrant an attempt at getting away, so I decided to give it a go. Everything was obscured by mist as soon as we got into the air, Bruay, being fairly high, had cleared before the surrounding country, but as far as the eye could see stretched miles of fog, extending from the ground up to about 4,000 feet and perfectly level on top. We were flying under a clear sky with nothing but a sea of what appeared to be white cotton wool. As far as looking the country over was concerned, it was hopeless but I led the formations on a south-east course to where I estimated Bapaume would lie, in the hope that the weather might be clear in that direction. Knowing that the country to the south was much lower and threaded by several rivers, I was doubtful even of being able to see the ground at all, yet alone study its geography. After about an hour I got fed up with the conditions and gave some thought as to where home was. Although I had flown on a steady course, I possessed no means by which I should know how far we had gone, other than by the amount of time we had been in the air. Some distance away I could see a row of observation balloons just poking though the mist and casually assumed that they were ours. As a matter of fact they were enemy balloons put up in preparation for the big offensive they were going to launch the next day, but at the time I had no means of knowing this.

We were flying north at 4,000 feet along the row of balloons and a little to their east, believing that we were on our side of the line, when suddenly three red Albatross scouts in "V" formation, poked their noses through the mist to the west of us. We were just above the surface of the fog and the sun had barely cleared it, and from their unconcern it seemed the glare must have prevented them from seeing us. They were followed by a string of other scouts of different types, all colored red, with the exception of one Albatross that was painted black and yellow — and they were flying on a course parallel to our own a little lower, and only a hundred yards away. As the black and yellow chap came opposite me I rocked my machine up and down, the usual signal to follow the leader into action, and turned and dived into the centre of them, choosing the black and yellow machine as my target. Courtney came down from above and for a while an all-in dogfight ensued. It was undoubtedly the Richthofen Circus, but the leaders had got into another layer of fog and missed what was going on. I managed to get my fellow in flames after a couple of bursts, and turned on to a red Albatross that was on the tail of Tab Pflaum. He turned to avoid me and then Robertson, who was chasing another scout, dropped dead onto him before he could get out of the way, putting a long burst into him and then chasing him down through the mist as he went out of control. In the meantime, Tab was having a busy time with two others and was concentrating more upon keeping out of their lines of fire than hitting back. I came up under the tail of one of them, a red triplane, and fired a burst from about ten feet away. He heeled over and went straight down through the mist all "arsey tarsey". Tab had meanwhile got onto the tail of his opponent and was firing straight into him, when they both disappeared from view below. My last sight of Elwyn, a new pilot, had shown three Huns trying to shoot him down some distance away, but he was putting up a good show and giving as good as he was getting. Elwyn had only joined us that month.

In a few moments it was all over, and Wright and I were the only ones of our formation left above the mist. I circled round for a few moments and presently Tab came pushing up through the murk, but there was no sign of Garth Elwyn or Robertson. I made for home, full of apprehension about the fate of our two missing pilots.

After a mile or so another Camel came out of the mist well over to the west and on approaching him I found that it was Robertson, apparently none the worse for his adventure, but there was no sign of Elwyn.

The Bruay 'drome was still fairly clear and we got in without any trouble. We had hardly reported the day's work when a ring came through from Elwyn. His engine had been badly shot up, but he had managed to cross the lines safely and land near Courcelles. He turned up later in the day and we all congratulated ourselves on the successful culmination of the flight's first "do" with the celebrated Circus. We admitted our great luck in that three enemy leaders missed the show; if they had joined in the fight an entirely different ending would most likely have ensued, especially as one of them was most likely Richthofen.

The Big German Push

CHAPTER 5

The next day, 21st March, the Germans opened their big push. Things began to happen quickly, and the squadron commenced the busiest period of its career. Fortunately we were at full strength and had a few pilots in reserve. Roy King, who was destined to become one of the outstanding pilots of the unit, had joined "A" Flight a couple of days previously, and although from his large stature he had some difficulty in fitting into a Camel, and found trouble in landing them because of this, he nevertheless started right off by taking his place in the flight. The C.O. who was impatient of mistakes made by other people, used to stand on the tarmac and curse every time "Bo" (as he was known) tried to make a proper landing with his tail down. Bo was well over six feet tall and broad as well, and the trouble was that he did not have enough room to pull the stick back and to get his tail down properly when at gliding speeds. In the air he was as good as gold, as far less movement was required to handle the controls, owing to the higher airspeed and the extra pressure on the flying surfaces. Anyhow for a time it was a toss up whether he would stay or not. Tab had known him during his training and had told me that he was good value, so I took it upon myself to ask that he be allotted to my flight and to become one of us.

Malley had now succeeded to the command of "B" Flight, and a number of other changes occurred. Our old pal Bill Brasch had been admitted to hospital seriously ill, and we said cheerio to him with many regrets. He was a sticker in all sorts of trouble and could always be relied upon to be on hand when you barged into a scrap, a state of affairs that was most satisfying to a leader.

Tab also departed for Blighty at this time, so two

Lieutenant Roy 'Bo' King above and, below, R.G. Smallwood, two of Cobby's closest friends. King destroyed 16 enemy aircraft and Smallwood three.

more of the original "A" team had gone. I missed Tab tremendously, and was now the only one left of the original flight. If we had not been so busy I would have felt lonely, even though the new chaps were first-class fellows.

From the 21st March onwards we were ordered to operate on the new front over the enemy "push" and the job consisted of getting to the line loaded up with bombs and ammunition, as fast and as often as one could, and letting the enemy on the ground have it as hot and as heavy as possible. The first sortie would commence well before daylight; the last would finish after dark. All this flying was done under 500 feet and our targets were point-blank ones. Frequently we would meet whole columns of troops on the roads and the havoc caused by our bombs and machine guns fire was tremendous. The air was full of aircraft, and continuously while shooting up the troops on the ground we would be attacked by enemy scouts. They would drop through the clouds and the mist, have a quick snap at us, and would then duck back into the clouds again. The weather was as bad as it could be with clouds and mist right down to within a few feet of the ground. The smoke of the battle mixed with the clouds and mist above, rendered flying particularly dangerous, quite apart from the extreme risk of running into both the enemy's and our own shells. On top of this there were the scores of machine guns devoting their time to making things as unpleasant for us as possible. We all averaged about six trips a day and this went on for nearly four weeks. The pace could not last and it was not long before we were all starting to show signs of the strain. My principal trouble was that I could not eat, but champagne and brandy with an odd biscuit seemed good enough. It was nothing more than overstrain, nervousness, and a mixture of fear and high tension. Added to this was the feeling that the war was going the wrong way and that our efforts were wasted. The line literally jumped forwards miles daily and there did not seem to be sufficient troops or guns to stop the German advance. In those first days we would come across bands of British troops who had not had a chance to get away and were still fighting, with the enemy miles beyond them. No one will ever write their final history. Owing to the presence of so much cloud and mist, machines were continually becoming separated from their formations, but as fast as three or four were ready to leave the ground on a fresh sortie, they would get away irrespective of which flight they belonged to. Our aerodrome soon became the base for a number of squadrons from other areas. Some had found it necessary to abandon their 'dromes in the face of the advancing German troops, and others were temporarily moved from their sectors. Some used Bruay because their own ground was still obscured by mist. We necessarily suffered many casualties, but they were not great compared to those of the Bosche. Our squadron in four weeks lost two pilots shot down and killed, two taken prisoner through their engines being damaged by fire from the ground plus a number wounded. Most of the rest of us had our machines shot to ribbons, and on one occasion my 'bus was so badly knocked about that the seat collapsed under me. A burst from the ground whilst turning at about twenty feet had gone through the fuselage and shot away the bearers of the wickerwork seat. Incidentally, the seat of my flying suit was ripped in three or four places by the selfsame bullets or some of the splinters of the bearers, and I only just managed to retain sufficient control to keep the machine straight by jamming my neck up against the back of the cockpit, stiffening my knees, and arching my back. I came home about thirty miles like this and landed by shutting off my machine and hanging on with one arm over the side of the fuselage. I finally came to rest the right way up after a few bounces. A somewhat similar thing happened to one of our visitors from No. 20 Squadron. They were flying two-seat Bristol Fighters, and one of them was so shot about that his machine collapsed as it landed. All the landing wires had been shot away and most of the controls. As it came to rest it just plopped down like a wounded duck. The crew refused to be worried by such a triviality and waited for the next of their machines to arrive, which it did just about dusk. The two climbed aboard with the pilot and observer who were already in the machine, and flew off about seventy miles north to their own aerodrome. They were one of the stoutest bunches of fellows it was ever my good fortune to meet. They were nearly all Canadians and used to make the trip to Bruay in the early morning darkness, do a heavy day's fighting from Bruay, and then fly the seventy odd miles home again in the late evening. Sometimes they put up in the mess on odd couches and whatever we could provide in the way of sleeping accommodation, but these nights were given somewhat to revelry and not much sleep was had by anyone. They knew more songs than any other crowd I had ever met and several could play the piano quite well. It was a noticeable thing that very few Australians could play any musical instrument, yet as a race we are as fond of music as anyone.

Just about this time we were provided with an improved engine, the designers having boosted up the compression by putting in longer connecting rods. The improvement was noticeable at higher altitudes and we were able to get to a greater height. This naturally meant a great improvement in performance.

On 26th March, the Germans were pushing out northwards and westwards from Bapaume and had extended their activities on the ground along the front before Arras to the north, and were spreading fanwise towards Amiens. The villages of Ervillers, Achiet-le-Grand and Miraumont were in their hands, and our immediate task was to do as much damage as we could to enemy troops in these places and along connecting roads. It was almost impossible at the height we were flying to miss targets with our bombs or machine-gun fire. The casualties must have been enormous, both on advanced troops and reserves. One only had to fly straight over a road for a few seconds, and let the bombs go one after another as quickly as possible, and they fell amongst masses of marching troops. Long afterwards it was ascertained that our aeroplane attacks affected severely, if not decisively, the success of the German offensive. His determination to push through to Abbeville and separate the British and French forces was frustrated largely from the air. An historical reference to this particular period states, "A thorough examination of the action of the Allied Forces at this critical time before Amiens, in the closing days of March, makes it almost certain that while it was the heroic infantry of outnumbered British and French Divisions which held up the advance of the enemy, and the Australian divisions played a glorious part in the later stages, it was principally the untiring exertions of the airmen in delaying, damaging and disheartening the enemy's reserves, and throwing his whole transport system out of gear, which enabled the Allied infantry to succeed."

We who took part in the show were perhaps overimpressed by the might of the enemy on the ground. He

Report on Low Flying and Bomb Dropping by 4th. Squadron, A.F.C. 21-4-1918.

Capt. W.B.Tunbridge	Sop. Camel	D1818	Dropped 4/25lb. Cooper Bombs on Lestrem from 2,000 ft at 8-30 a.m. Fired 200 rounds at Troops near LaPierre from 2,000 ft.
Lieut. N.C.Trescowthick	do.	D1838	Dropped 4/25lb. Cooper Bombs on Lestrem from 2,000 ft. at 8-30 a.m.
Lieut. R.G.Smallwood	do.	B2535	Dropped 2/25lb. Cooper Bombs on Lestrem from 2,000 ft. at 8-30 a.m. Fired 200 rounds at Troops on road near Paradis from 2,000 ft.
Lieut. G.F.Peirce	do.	B2489	Dropped 2/25lb. Cooper Bombs on Lestrem from 2,000 ft. at 8-30 a.m.
Lieut. P.K.Schafer MM	do.	D1863	Dropped 2/25lb. Cooper Bombs on LaGorgue from 5,000 ft. at 11-30 a.m.
Lieut. E.R.Jeffree	do.	B8231	Dropped 2/25lb. Cooper Bombs on LaGorgue from 5,000 ft. at 11-30 a.m.
Lieut. A.H.Cobby	do.	C8226	Dropped 2/25lb. Cooper Bombs on LaGorgue from 5,000 ft. at 11-30 a.m.
Capt. W.B.Tunbridge	do.	D1818	Dropped 4/25lb. Cooper Bombs on LaBassee-Lille Rd. from 2,000 ft. at 7-20 p.m.
Lieut. R.G.Smallwood	do.	B2535	Dropped 2/25lb. Cooper Bombs just East of LaBassee on the LaBassee-Estaires Rd. from 2,000 ft. at 7-20 p.m.
Lieut. N.C.Trescowthick	do.	D1838	Dropped 4/25lb. Cooper Bombs on LaBassee-Crassiers Rd. from 2,000 ft. at 7-20 p.m. Fired 100 rounds at Troops along the LaBassee-Lille Rd. from 2,000 ft.
Lieut. R.King	do.	B2527	Dropped 2/25lb. Cooper Bombs on LaBassee-Lille Rd. from 2,000 ft. at 7-20 p.m. Observed direct hit with one bomb dropped by Capt. Tunbridge. Fired 150 rounds at Troops along same road from 2,000 ft.
Lieut. G.F.Peirce	do.	B2489	Dropped 2/25lb. Cooper Bombs on LaBassee-Lille Rd. from 2,000 ft. at 7-20 p.m. Fired 100 rounds into support Trenches West of LaBassee from 2,000 ft.

R.R.B. Bennett
Major.
Commanding 4th. Squadron, A.F.C.

Reproduced here is a sheet from the Official War Diary of No. 4 Squadron, A.F.C. It lists part of the intense activity of the time. The signature is that of Lieutenant R.R.B. Bennett, the squadron's recording officer. At left is Captain Wally B. Tunbridge, the commander of "A" Flight before Cobby succeeded him.

seemed to possess such vast numbers of men, and it appeared that no matter how we threw ourselves into the fray, the effect of our efforts were so small that it did not seem as if they could produce any result. Of course, our knowledge of what was going on was confined almost exclusively to what we, as individuals, were concerned with. Naturally we were aware that many chaps in the air were doing the same thing all round us and behind the lines, but our efforts seemed so puny, even though we were inflicting tremendous damage to personnel. It was ruthless and relentless warfare, and most hazardous. Looking at this period in retrospect, one realizes that our concerted work in the air prevented the sending forward by the enemy of foodstuffs and ammunition to his advanced troops. We knew that very few wheeled vehicles were able to move on the roads either side of Bapaume, but being so keenly concerned with our own particular job, we lost sight of the complete picture. Of course we were only doing a fraction of the total work, although we probably had an idea at the time that we were doing a lot. I know it seemed like that, but later on we realized that every pilot and machine that could be spared was thrown into the battle with instructions to count no cost whatever, but to fling himself at the mass of the German divisions behind the advanced troops, to smash it and go on smashing it night and day. It was the beginning of the Allied victory, although we did not realize it at the time. In fact our views were then almost the opposite.

No sooner had things quietened down on the Somme towards the end of March, than a fresh attack was launched by the enemy further north. The earlier advance was definitely held in front of Amiens in the vicinity of the line of Corbie, Villiers-Bretonneux and Albert. Reinforcements, including the Australian Corps and a French Cavalry Division, had arrived at the right place at the right time, just when he had over-reached himself. It seemed, however, that Germany had determined to maintain the offensive as the only means of winning the war. The new attack was opened between Bethune and Ypres, opposite a portion of the line held by the Portugese. The new area was part of our old hunting ground and was much nearer home. Only forty minutes or so was required to get out, return and load bombs again, and away we went at the same hectic pace. This time Bruay seemed to be on the intended line of advance, as all signs pointed to an attempt to reach Bethune. Preparations for this attack were obviously being made near La Bassée, and this was going to bring the war to a successful close. So the La Bassée-Lille Road was our first objective.

It was about 9th April when we made the switch back to our old happy hunting ground from Lens to Ypres. One of the first patrols was a squadron affair, when we were ordered to bomb the La Bassée-Lille Road. We had barely started over the line when we met with murderous anti-aircraft fire. We were in a squadron formation in "V" shape, with my flight on the left. Malley's was on the right and "C" Flight leading in the centre but a little lower. I glanced quickly at "C" Flight, and suddenly saw the leader of the second three go into a vertical dive. Being an unusual occurrence for a member of a flight to leave the formation in such a manner, I tried to see what he was diving at, but could see nothing. I then noticed that there was something a few feet in front of the engine that looked as though it might be part of the engine itself, but while I

(Please turn to Page 57)

"A" Flight on 16th June 1918. Cobby is in the centre with Smallwood on his left in the photograph and King on his right. The names of the others are unknown. Note the range of attire, Sidcot suits for some and leather coats for others. The 'Sidcots' were designed by an Australian, Sidney Cotton. The one-piece multi-layered flying suit was to prove of immense benefit to pilots for a quarter of a century to come. The same designer pioneered the anti-gravity flying suit.

© Norman Clifford

THE BE2A
From 1914 until 1918 two of these aircraft were employed as advanced trainers with the Central Flying School at Point Cook, Victoria.
Dimensions: Span 38 ft. 7½ in. Length 29 ft. 6½ in.
Power unit: 70 h.p. Renault stationary air-cooled engine.
Speed: Between 60 and 70 m.p.h. at sea level.

THE BRISTOL SCOUT D-TYPE
This aircraft flew only twice during its life at Point Cook. It was considered too dangerous for graduating pilots to fly.
Dimensions: Span 24 ft. 7 in. Length 20 ft. 8 in. Height 8 ft. 6 in.
Power Unit: 100 h.p. Monosoupape nine-cylinder air-cooled radial rotary engine.
Speed: Approximately 100 m.p.h.

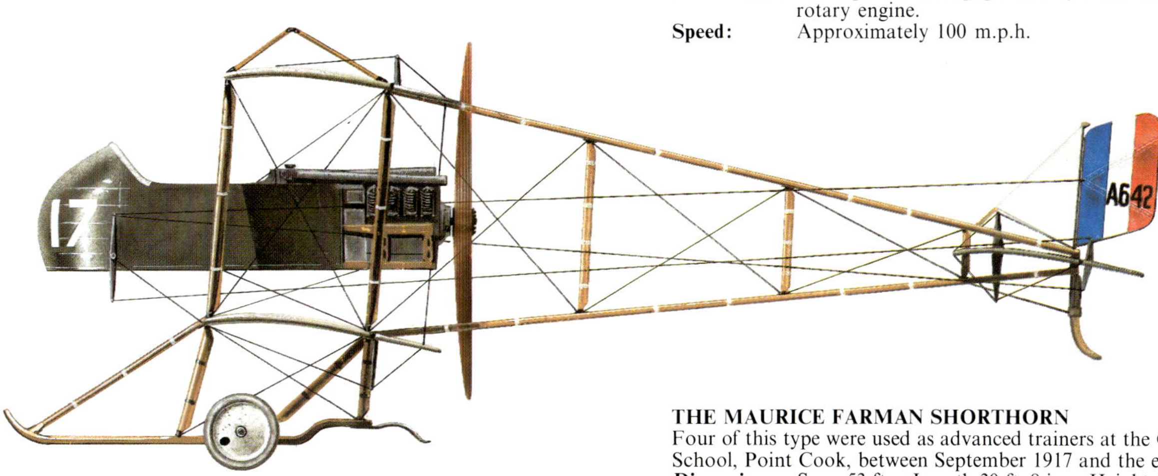

THE MAURICE FARMAN SHORTHORN
Four of this type were used as advanced trainers at the Central Flying School, Point Cook, between September 1917 and the end of the war.
Dimensions: Span 53 ft. Length 30 ft. 8 in. Height 10 ft. 4 in.
Power unit: 80 h.p. Renault eight-cylinder stationary air-cooled engine.
Speed: Between 60 and 70 m.p.h.

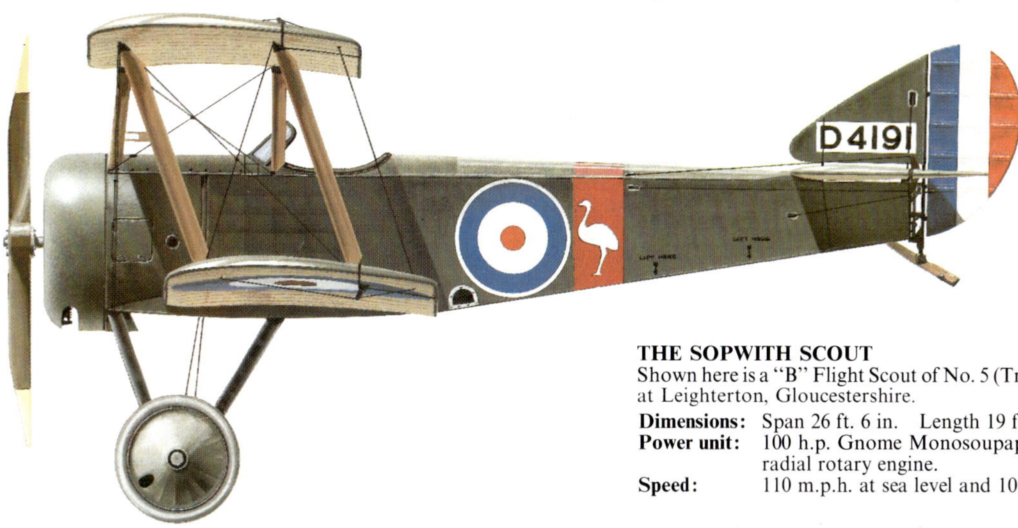

THE SOPWITH SCOUT
Shown here is a "B" Flight Scout of No. 5 (Training) Squadron, A.F.C., at Leighterton, Gloucestershire.
Dimensions: Span 26 ft. 6 in. Length 19 ft. 3¾ in. Height 9 ft. 5 in.
Power unit: 100 h.p. Gnome Monosoupape nine-cylinder air-cooled radial rotary engine.
Speed: 110 m.p.h. at sea level and 105 m.p.h. at 10,000 ft.

©

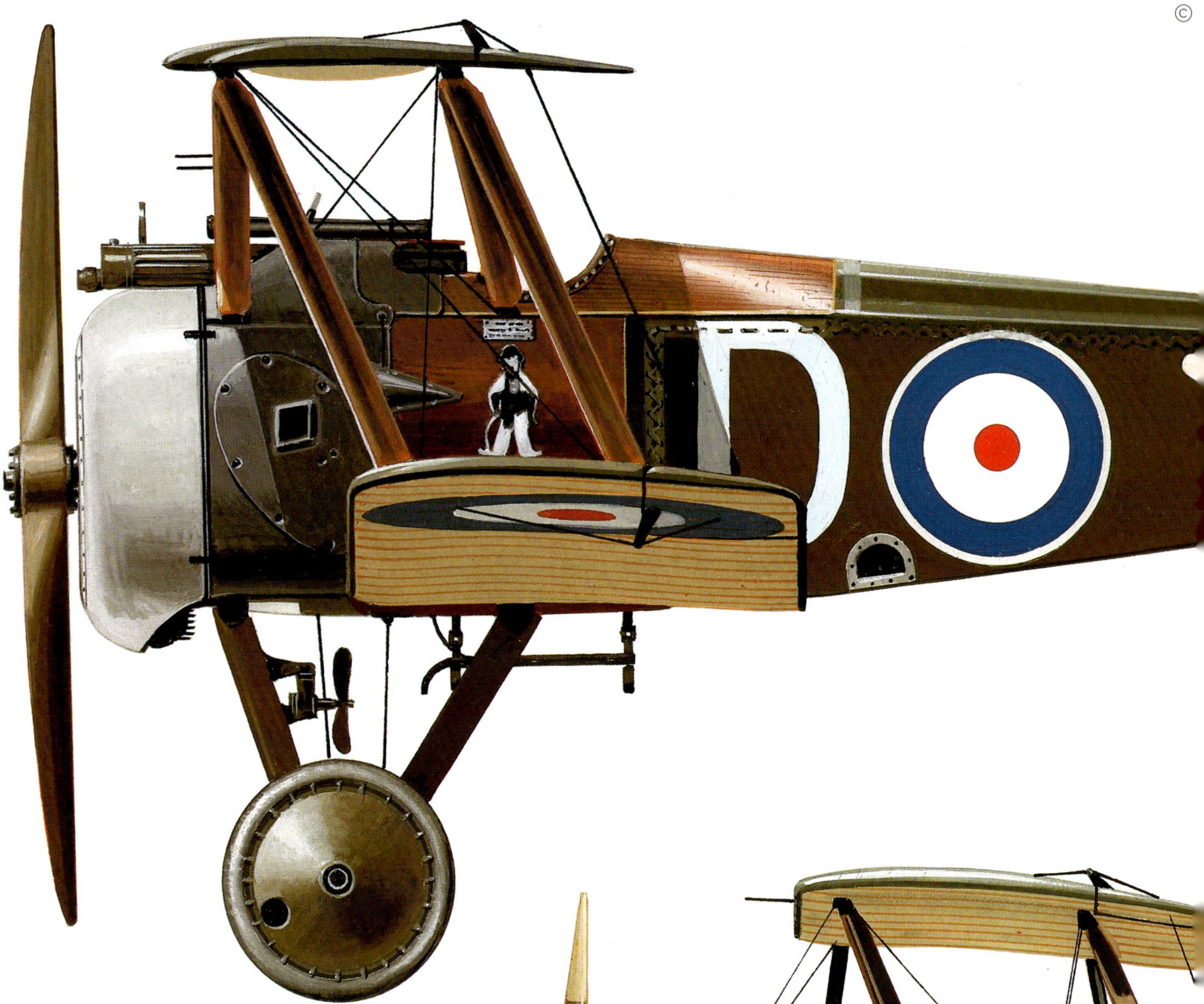

HIGH ADVENTURE

Published by
Kookaburra Technical Publications Pty. Ltd.

Clifford

THE SOPWITH CAMEL F.1

This machine was flown at Bruay by Captain A.H. Cobby early in his term with No. 4 Squadron A.F.C. The object under the fuselage between the rear wheel struts is a rack for Cooper bombs, mounted four abreast.

Dimensions: Span 26 ft. 11 in. Length 18 ft. 6 in. Height 9 ft. 1 in.
Power unit: 130 h.p. Clerget (short stroke) nine-cylinder air-cooled radial rotary engine.
Speed: 115 m.p.h. at 6,000 ft. and 109 m.p.h. at 15,000 ft.
Service ceiling: 22,000 ft.
Armament: Two forward-firing synchronised .303 Vickers machine guns.

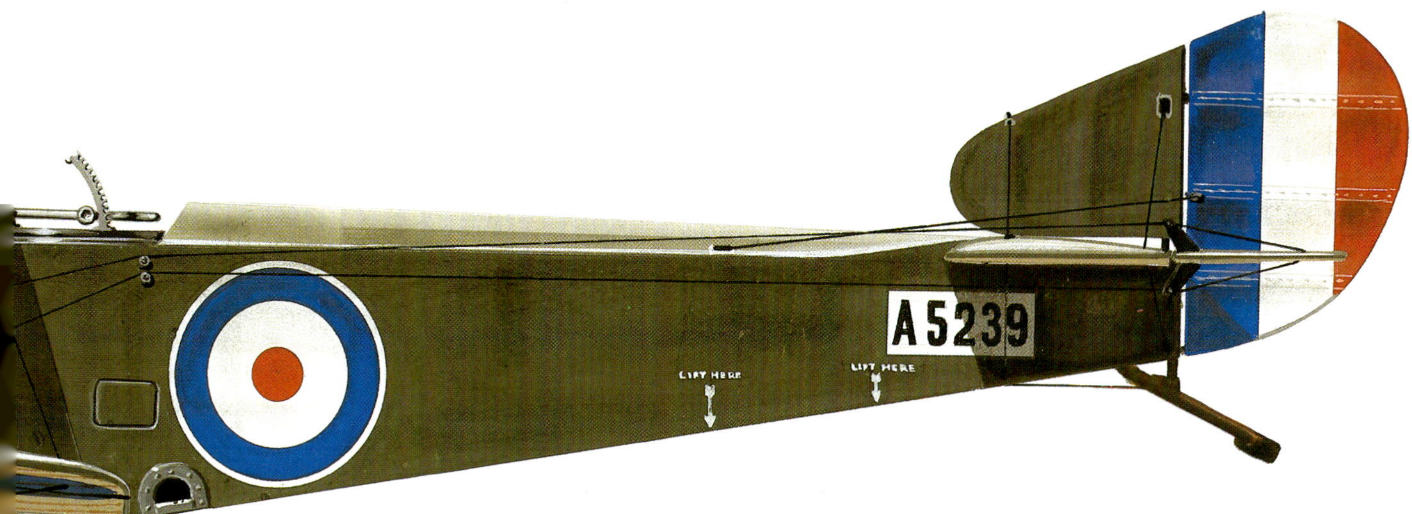

THE SOPWITH TWO-SEATER

This was a front-line service aircraft during the first half of the war and a training type for the remainder. Shown here is a machine of No. 68 (Australian) Squadron, R.F.C., later to become known as No. 2 Squadron, A.F.C. The Two-Seater was considered an important link between elementary and advanced training types, particularly for trainees graduating to rotary-engined fighters.

Dimensions: Span 33 ft. 6 in. Length 25 ft. 3 in. Height 10 ft. 3 in.
Power unit: 110 h.p. Clerget nine-cylinder air-cooled radial rotary engine.
Speed: 103 m.p.h. at 4,000 ft. and 92 m.p.h. at 12,000 ft.

THE SE5A
This was the personal "C" Flight aircraft of Captain Les Holden of No. 6 (Training) Squadron, A.F.C., at Minchinhampton, Gloucestershire.

Dimensions: Span 26 ft. 7½ in. Length 20 ft. 11 in. Height 9 ft. 6 in.
Power unit: (Aircraft shown), 200 h.p. Hispano-Suiza (geared) V-8 stationary water-cooled engine.
Speed: 121 m.p.h. at 15,000 ft.
Service ceiling: 17,000 ft.
Armament: None on training aircraft.

THE AVRO 504K
The aircraft depicted here is one flown in No. 5 (Training) Squadron, A.F.C. at Leighterton, Gloucestershire, by Captain "Tabby" Pflaum.

Dimensions: Span 36 ft. Length 29 ft. 5 in. Height 10 ft. 5 in.
Power unit: (Aircraft shown), 100 h.p. Gnome Monosoupape nine-cylinder air-cooled radial rotary engine.
Speed: 62 m.p.h. at 8,000 ft. and 75 m.p.h. at 10,000 ft.
Service ceiling: 13,000 ft.

THE SOPWITH CAMEL F.1
This was Captain A.H. Cobby's distinctively marked "C" Flight Camel whilst he was Wing Fighting Instructor with No. 5 (Training) Squadron, A.F.C. at Leighterton, Gloucestershire. This aircraft is referred to in Cobby's story.

THE SOPWITH CAMEL F.1
Another easily identified Camel was this one flown by Captain Garnett Malley whilst he was "C" Flight Fighting Instructor with No. 6 (Training) Squadron A.F.C. at Minchinhampton, Gloucestershire. Like Cobby, he was previously a leading pilot with No. 4 Squadron, A.F.C.

was endeavoring to identify it, the machine burst into flames. There was nothing to indicate what had caused the machine to dive in the first place, but it had most likely taken a direct hit from an unexploded shell on its way over, and had smashed the Camel to pieces. I knew from the position in which he was flying that it was Courtney. There was practically nothing left of his machine by the time it reached the ground. The new war was much nearer to Bruay than the one we had left behind on the Somme, and consequently our trips were more frequent. It was again our job to try and take the place of the guns that had to be withdrawn owing to the rapid advance of the enemy, and to interfere as much as possible with his movements. The weather was still misty but was not so thick as during the previous few weeks, and it gradually improved. We were unfortunate in losing several pilots about this time. One of them happened to be again our best performer on the piano, a chap named Storch. We had taken particular care to see that he was well guarded on all trips over the line, and had informed him that his playing of the piano was almost as important to us as the ability to shoot down Huns. He considered that he was lucky to be so well cared for. Unfortunately the best laid plans don't always work out as they are intended to and in this case they definitely did not. We had returned from a stunt over the line and most of us had landed, but Storch, being new, always put in some time on the ground target and commenced a series of dives to practise with his guns. In one approach he dived too steeply and had to jerk his machine out suddenly, but left his wings behind. He struck the ground and was killed instantly.

Another one of our tasks in this enemy offensive was the bombing of the crossroads between Neuve-Chapelle and Estaires, and moving columns of guns and transports on the many roads that approached the line in the vicinity, especially near Richebourg. S'Vaast and Laventie. The short distance we had to fly enabled our sorties to be made much more rapidly than in the previous "do" and the number of trips we could make in a day naturally increased our hitting power. Again, the order of the day was to get out and back again as quickly as possible and it was ding-dong from before dawn until after dark. Again it was point-blank warfare. It was a case of every time you saw a head you hit it, and there were plenty of heads. Bombs and bullets would create havoc among formations and before they could reform, another relay of machines would arrive and smash them again. The bad weather made little difference to the effectiveness of our attacks, as we only had to fly in the clouds for about ten or fifteen minutes, poke our noses down, and there were always plenty of things to have a go at. We caused tremendous havoc in the around Merville, and anything that was able to get past that hectic spot on the ground was lucky.

One rather interesting event took place in the air on the morning of the 10th. We had not contacted many enemy aircraft up to that date, but just as we approached La Bassée in flight formation at about 2,000 feet, an Albatross scout came out of the clouds and proceeded to fly alongside us. I had not noticed its arrival as it had most likely just emerged from cloud, and I was on the point of going down through the cloud bank when Bo King edged up alongside me and with many gestures pointed it out. This was one of King's first trips, and I gave him a thumb sign to get after the Albatross and pulled up after him. The Hun moved into a cloud and King went over the top to catch him as he came out on the other side, but I went round the cloud to the right and met the other chap almost head-on. There was just time to press both triggers and to dive under him to get out of his way. He went down into the clouds in flames, but I had hit the pilot, and he almost jumped backwards out of the cockpit when I fired. Bo was adrift for a while, but caught us up later.

Merville, St. Venant, Roberque, Caloune and Pacant were attacked from point-blank range continuously on the 12th. Just short of the village of Aaradis, we came upon a large column of troops marching north to Merville at about 3 p.m. and the six of us immediately went into line ahead formation and dropped our bombs along the column. Everyone of the twenty-four landed on the roadway, and as the troops who were left took to the fields, we turned and machine-gunned them. They were useless as a co-ordinated force when it was over, and it would be some time before they stopped shivering when an aeroplane came near them.

The enemy did not progress far beyond Merville, as on 15th and 16th April, strong reinforcements arrived for to the sorely pressed troops, which included the first Australian Division. No. 4 Squadron then concentrated on the well held German positions in front of Bethune. We plastered all this area on 18th, 19th and 20th and then switched for a few days to the villages of Lestrem and La Gorgue, which contained many dumps and gun emplacements. Things again settled down to normal by the last week in April, and we resumed our normal offensive patrols.

One most interesting incident occurred during this period which illustrates the splendid spirit of the fellows who were flying. I had been over the line with a formation, and had arranged that when the job was finished, I should leave them and fly back and see if I could pick up a sight of any British or French reinforcements that might be moving up. This was necessary as ground movements were fairly rapid and we did not want to risk attacking our own troops. Further, it was easy to confuse the German field grey uniforms with those of the French and Portuguese.

Some distance back I noticed some heavy guns being moved into position, and from the direction they were pointing they were obviously ours, so after circling around low down for a while I landed in a field nearby and went down to them. Whilst conversing with the British battery commander, another Camel landed in the same field and came to a stop, but no one got out of it. We went across and in the cockpit a very young English lad was trying to force the watch off the instrument panel with a pocket knife. On enquiring what the idea was, he said that he had finished with the war, and wanted to take a souvenir home. He said he had only been in France a few days and it looked as though he would not be back for some time at least. On asking him what it was all about, he said he had been in a scrap and wounded in three places and was not going home without something for his mother. So we helped him out and looked him over and found he had been shot high up through the right chest, and had a bullet through each leg. None seemed particularly dangerous, but he refused to be moved until we souvenired the watch for him. We then had him patched up and sent away. Some weeks later I received a letter from him and another from his mother. He was doing well but it seemed that his shoulder was going to suffer some limitation of action, but he was not very concerned about that. He must have only been about 18 years old.

About this time we were subjected to a taste of what

the troops in the line and in reserve perhaps accepted as an everyday occurrence, but to us it savored of unfair tactics by the enemy. The enemy started to shell us. It was very disturbing and was responsible for a lot of sleeplessness and blasphemy and brought to us a phase of warfare with which most of us were unfamiliar. It was pretty big stuff he was putting over, probably about 11-inch shells and for periods of about half an hour several times during the day and night there would be "crump!!!" and a bang every four minutes. Nissen huts would rock, fragments of high explosive shell would whistle and scream through the air and bricks from a nearby terrace of houses would thump around us, even though no shell fell closer than about thirty yards away. Of course in our conceit, we assumed that the squadron was the target, but it is for more likely that the railway line to Bruay that carried all the coal from the local mines was his real objective. This line ran along just at the rear of our huts, and for the whole period of the bombardment was not seriously damaged, but a long terrace of the usual type of French houses about thirty yards further away and running parallel to the railway, was reduced to practically nothing. It was almost perfect shooting. Tab Pflaum used to say that one of the gun crew must have been leaning up against the barrel of the gun, but as the gun was probably about forty or fifty tons weight, this can be regarded as a piece of Tab's humor. At any rate, considering the distance that it must have been away, a most minute fraction of an inch alteration to the line of the gun would have meant the demolition of the railway or the aerodrome. Admitting that while the fire was so accurate we were in comparative safety, we nevertheless were no longer comfortable on the aerodrome, and tried every means available to discover the location of the gun so as to be able to retaliate, but we were unsuccessful. Good luck to them anyhow, they did a lot to spoil life for us at the time, and if their receptive faculties were able to pick up the things we said about them I am sure those German gunners would have been satisfied that they were doing their bit for the Fatherland.

Bethune was bombed for the first time towards the close of the enemy northern push. It was apparently realized by the German High Command that there was little hope of them capturing the valuable area containing the coalfields, so right out of the blue the unsuspecting populace was subjected to an intense air raid. Unfortunately some of us were in it too.

At this time the town had grown to many times its normal size as many people in the nearby war zone had drifted into the sanctuary of Bethune. For two or three years it had been regarded as a safe area, and although almost in sight of the line its social life had been fairly normal. Busy streets, well attended markets and an unusually large shop trade were ordinary affairs of the day, when without warning, several waves of enemy aircraft bombed and machine gunned the town. From the quiet old world atmosphere of a large provincial town, everything changed in a few moments. The market square was the centre of the attack, and it was market day. A peaceful scene was immediately reduced to a state of indescribable chaos. Horses and cattle broke loose, citizens and soldiers were struck down, and houses and shops were wrecked.

Several of "A" Flight were shopping in the town at the moment of the attack, and I have never felt so impotent. There was no means of hitting back and you just had to stand and take it. All you could do was to keep out of the way of flying splinters and debris, and trust that a direct hit would not come your way.

This attack only lasted about half an hour, but the damage was extensive. None of us was injured, but on making our way back to the car, three of us fell through a skylight into the basement of a house that had been hit. We all suffered bruises and scratches, but what was more serious, Tab lost his silver flask. This happened to be a very fond and popular possession, having been presented by some local person at home when he left Australia, and it had performed yeoman service. All our searching failed to find it. It used to hold about two-thirds of a bottle and had accompanied Tab on all his operations, both warlike and peaceful, in various parts of the world. For many months I had seen him tilt it to his lips on the way home from patrols and on other occasions. From then on he always looked for a pilot's mess adjacent to any enemy 'dromes we strafed, and made a personal attack on it.

I was now regularly leading the flight, and three or four times a day we performed the short pilgrimage leading into enemy territory on ground strafing duty. Only an odd scrap or two came our way during the remaining days of April, and things were becoming stale. I think that we knew every cobblestone in and around Estaires, Croix Barbée, Rouge Croix, Richebourg, Neuve Chappelle, Lacon, Zoboles, Le Cornet, Malo and Lens, and all the interesting roads in this district.

For some time Malley, Tunbridge and I had been the only remaining pilots of the original batch, and we had had a pretty severe time of it. Some of the chaps had been on leave to Blighty and had come back more or less rejuvenated and my turn now came along. I now had visions of fourteen days and fifteen nights in Blighty!

I was warned that I could start my leave in a couple of days' time and you can imagine how careful I was not to expose myself to undue danger during this short period. After borrowing what odds and ends of clothing I could, collars, ties, shorts and so on, and raising a loan or two from the chaps, off I went to Boulogne, en route for the Old Country. What a wonderful feeling of peace of mind and contentment in the realisation that for the next couple of weeks I would be free of the war! For fourteen days I would not have to creep out of bed at the unearthly hour of two or three in the morning and lead my team across the line into sticky situations, nor would I have to exercise my mind as to how I could extricate them. Baths, fresh linen, English food and the English language were to be my lot for the time being and the thought was all sufficing. I was much more fortunate than most "colored troops", as Dominion armed citizens were known, in the fact that I had several lots of relations in and around London. The difficulty of making one's pay meet the cost of activities that one desired to indulge in was much easier for me, as hotel bills were things that I did not have to worry about. Quite a number of fivers and tenners also came my way from an aunt at whose place I generally stayed. We were always most welcome and my aunt was a second mother to my brothers and I any time we could get to London.

This first leave was a quiet affair. Theatres and the river were my principal sources of enjoyment, and both being handy, my evenings were spent in seing every show that I could get to, and my afternoon in sculling on the Thames. I regarded with scorn the manipulator of a punt, but I was to learn that there was a lot to commend this form of enjoyment. Some friends of my cousin's arranged a picnic in punts, with a rendezvous at one of those delightful little islands that dot the river up beyond Hampton

Court. We were paired off and I was surprised at the number of cushions that it appeared necessary to pile on board. I disdained to use a "pole" but applied a paddle to the job, and after tacking and yawing backwards and forwards for an hour or so against the stream, I arrived at our destination. I now realized the advantages of a punt and the significance of the cushions. The long flat bottom and the heap of cushions now proclaimed their utility, and if anyone knows anything more pleasant than being in a punt tied under the willows, "far from the madding crowd", with a pleasant young woman beside one and an appropriate lunch basket handy, then I am prepared to try it at any time. Omar knew what he was talking about when he wrote,

"Here with a loaf of bread beneath the bough,
A flask of wine, book of verse and Thou
Beside me singing in the wilderness
And wilderness is Paradise now."

Yes! There's a lot to commend a punt!

Our Old Battle Front

The famous Fokker Dr. I Triplane was designed expressly as a "dog-fighting" machine and so manoeuvrable that it was difficult to hit. Dominant when it entered service, it became more vulnerable as the Allies introduced better aeroplanes. It was in this type that Richthofen eventually lost his life after destroying his 8oth victim.

CHAPTER 6

In due course I arrived back in France and was met at Boulogne by Malley in the squadron touring car. He was full of news of big fights on the line and squadron gossip. Where a few weeks before it was difficult to make the enemy come to grips in the air, he was now coming out into the open and several decent dogfights had taken place.

At 7.15 on the morning of 15th April, I led my first patrol over the line following my leave. There were two flights, "A" and "C" with five machines in each and our job was offensive patrol on our old battlefront over and beyond Arras and Armentieres. We saw many large enemy formations during the patrol. One one occasion there were about twenty Albatross and Pfalz scouts to the east and well above us, and later a large body of the same sort of machines dived on us, but they would not come too close. We could not get to their height and the odds were too great for us to try and attack them from below, so I judiciously led the patrol out of danger. Numbers of other enemy machines were seen to the east of Bailleul, but they would not attack, so after dropping our bombs along the Lille road we returned home. Later that afternoon I led off a patrol of six and chased a batch of five Pfalz and ten Albatross scouts, but they would not give battle despite their superior numbers and as they had the legs of us, they got away. We vented our tempers on three observation balloons at the back of La Gorgue and tried to burn them, but they would not catch fire, so we gave that up too, after they were pulled down. We finished up by dropping our bombs on the landing ground at La Gorgue in the hope that we would put a few holes in the ground and cause some accidents to the Bosche planes using it. Later again in the evening I led eleven of our fellows out on a show, but other than seeing some seventeen or more of the enemy, some distance away, there was nothing to report.

We could not understand the attitude of mind of the leaders of all these enemy formations we were seeing. In every case they were in vastly superior numbers to ourselves yet would not fight. Many insulting messages were dropped on their aerodromes in an endeavor to force the fighting, but this was of no avail. We even went out singly and in pairs in the early mornings and shot up their aerodromes with incendiary ammunition to try and destroy their aircraft in their hangars. We did get some results from this as we managed to burn several aircraft waiting on the tarmac, but in the air they were still shy.

One method of drawing the attention of enemy pilots to the fact that they seemed to be staying at home a lot was to drop a pair of old boots in the centre of an aerodrome while shooting it up. This was not an original idea of ours as a Major Dallas, a New Zealander who commanded No. 40 Squadron had previously used it. Such comments as "Boots — pairs — one Bosche pilots for use of" and "Boots heavy type — German airmen for the use of whilst staying on the ground" and so on were frequently employed. Other much more personal messages were tried, but of a nature that could not be stated publicly. One that was commonly used was to the effect that they represented a severe kick in the pants for the German Air Force, to such a degree that it had been booted out of the sky. One of the "A" Flight pilots improved on this. He made a parachute and attached to it an ordinary bedroom utensil of convenience and included a message suggesting it might come in handy if the pilots of the particular aerodrome upon which it was dropped were confined to their beds. He followed this procedure on several occasions.

When we dropped the boots we used to attach red, white and blue streamers or an Australian flag so that it would attract attention. We always added that the gift was from Australia. I have no idea how these facetious remarks or the presents were received, but I expect they would be frightened of time bombs being concealed in them, and they would be kept at a safe distance for days. There was probably a purpose behind it all, of course, but we could not see it. We then tried the old game of putting out a decoy, that is the sending out of one of our chaps well over the line with the object of kidding the enemy to come down and have a go at the "poor innocent." To do this successfully the whole thing had to be extremely well arranged and times and everything else well co-ordinated, otherwise it was possible to lose the decoy without being able to help him. The protecting machines would sit up high towards the sun and watch the sky in all directions while the decoy stunted around aimlessly well over into Hunland. His job was not a pleasant one, as he had to stick there until the enemy were well committed in a dive before he cleared out. The protectors would wait until they were in a position to cut across the attackers and then would drop on them like a bolt from the blue, leaving the decoy to look after himself or return to the fight if it lasted long enough. In this manner we accounted for a few of the enemy, but they could dive faster than our Camels. Unless we got close to them early in their dive, they would just keep on diving and so get away. It was good fun though, and kept our fellows' tails well up. If we were only able to encourage the enemy to get into a dogfight things were easy, as a Camel could outmanoeuvre anything on the line. At the moment, however, this did not seem possible.

For the next week or so we continued with fairly large formations, sighting the enemy in numbers without being able to get close enough to have a go at them, and having to be content with forcing down balloons and in annoying people in billets and back areas. Then orders were issued by the Wing that patrols should be enlarged by the formation of "circuses". Several squadrons would rendezvous over a given place, and at varying heights, with instructions to sweep well over the line and endeavor to catch the Bosche in our net. This went on for days but without result. Our formations were so large (there would be perhaps forty or so machines in the sweep) that the enemy patrols could see us from a long way off, and just removed themselves from danger. Although we were achieving the purpose for which the fighting squadrons were employed, namely, keeping the sky free for our army machines to carry on with the job of reconnaissance and photography, it was not to our way of thinking — war. We certainly played merry Hades with everything that we could find on the ground and roads, and life must have been chock full of incident for the poor unfortunates who had to exist in billets and villages behind the line. In this at least we must have impressed the enemy that he had by no means won the war. I often wondered what kind of an answer the German pilots used to give to any of these people, when asked what they were doing to stop us from having the free run of their teritory.

Several new chaps joined the flight during April, A.D. Pate, G.R. Pierce and W.A. Armstrong of the Royal Air Force and R.G. Baxlain of the Australian Flying Corps. Poor old Baxlain did not get much of a go. On taking off he stalled his machine and spun into the trees at the bottom of the 'drome. He smashed himself up a bit and pushed his face in amongst his gun butts, and the next time I saw him, months later, his features were entirely different, but no worse to gaze upon. Garth Elwyn and Robertson had to be packed off to hospital and we had gone through two complete lots of pilots, and still I was going along, though somewhat apprehensive that it could not last for ever.

The squadron was now moved to Clairmarais on 28th April, a site some forty or fifty miles to the north, but we still operated on the same front. We were much nearer the coastal towns at the new 'drome, but it was with mixed feelings that we said goodbye to Bruay. Malley, Gillis-Watson, Bo Lamplough and others of us regretted the loss of the amenities we had become accustomed to at "Madame's" and both she and Leonie shed tears over us before we left. Nevertheless we were all quite willing to break new ground and looked forward to more frequent visits to St. Omer and such places as Boulogne, Calais, Etaples, Berck Plage and so on.

So we said goodbye to the 10th Wing and joined the 11th. We were now with the second Army under General Plumer. There was a good team of chaps in No. 74 Squadron, R.A.F., alongside us at Clairmairais under the command of Major "Grid" Caldwell, a New Zealander, with "Taffy" Jones, Carlin and others, but the outstanding personality was Mick Mannock, of imperishable memory to all who ever came in contact with him. Mick was the only man whom I met during the war who was 100% plus against the enemy. He was ruthless in his insistence that by any means within his power the enemy should be destroyed and he devoted all his energies and vivid mentality to this end. He was the inspiration and directing force in his squadron and was largely responsible through his advice and guidance for the maintenance of the standard which we ourselves had attained. In his determination that the only good enemy was a dead one, nothing was too much for him to take on. Whether it was by suggestion as to methods of fighting, demonstrations of tactics, or even the taking under his wing (literally) of someone not possessing his dash and verve, he gave himself wholeheartedly to the task and his enthusiasm was infectious. One couldn't associate with Mick for long without absorbing some of his boldness. Yet with it all he was a bright and whimsically cheerful chap.

He seemed to be able to play most musical instruments, and could improvise words to any topical song at a

A pilot's view over a Camel's twin Vickers guns. The aircraft in this case was Cobby's. The fitting of hump fairings over the guns was responsible for the name Camel. There are three sets of sights visible in this photo; the Aldis telescopic sight, on which was mounted a ring rear sight and a bead front sight. On the right hand Vickers was a bead rear sight and a ring front sight. Lower. No. 4 Squadron, A.F.C. shared Clairmarais aerodrome with No. 74 Squadron Royal Air Force, flying SE5As. One of their number at this stage was Major Mick Mannock, a V.C. winner who, before his death, had notched up 73 victories.

moment's notice. His main theme in his songs was a confusion to the enemy and I think Mick committed it to nearly every popular tune of the time. He also used to sing of himself going down *en flamme* at the hands of some enterprising young German, and was convinced that that would be his ending. His Irish nature even enjoyed talking of this. He must have been fey or psychic as a few months later he was shot down in flames on the line after a successful combat. His cobber Taffy went berserk for a few days and lived over the lines trying to avenge his passing, Mick would have been delighted if he could have observed the results, but I know that he had no regrets. He went as he always hoped he would.

Amongst other things at Clairmarais, we were given the job of interception of high flying enemy artillery machines. It proved somewhat abortive, as we could not reach them, but now and again we fluked contact. It was all worked from a ground directional wireless and we would have a couple of machines and pilots standing by for a call. It was quite a simple arrangement. Several of these stations would be connected by telephone, and as soon as an enemy machine started to register his artillery by wireless signal these stations would estimate where the machine was "in space". They would pass this information to us and the stand-by pilot would hop off and try and catch him. Sad to relate the Germans used to fly so high that we could not reach them and after falling out of the sky in a series of spins, we would let fire at long range, hoping for the best. Sometimes we would get onto one at

our operating height and shake it up, but I am afraid this seldom occurred. Even when well above our height, they usually cleared off, so in many respects the effort was worthwhile.

On 14th May Watson and I led our lights out to bomb enemy field artillery near Bailleul, but air combat were few and far between. Our interests were then directed towards annoying the enemy on the ground, and this was practically all we could do for several days.

On 20th May I led nine of our chaps on a more satisfactory job. We started off at seven in the morning on the usual offensive and bombing patrols and climbed towards Armentieres into a clear sky. There was a bank of cloud some distance over the lines, and I made across this in the hope that there might be some pickings on the other side which had not seen us coming along, There were! A couple of Pfalz scouts were idling along just on the lower edge of the cloud, some thousand feet below us, but about five thousand feet or so above was a nice little crowd of eighteen other enemy scouts. The temptation was too great to withstand, and having in mind the knowledge that the enemy had been so cautious recently, I rocked my machine to indicate that something was going to happen, and then kicked it into a half roll and dropped on to them in a vertical dive. Pate was alongside me almost as soon as I started and he went for the one at the rear while I aimed at the forward one. One burst from about fifty yards was enough to put my opponent into an uncontrolled spin, so I pulled out of the dive and turned to give Pate a hand with his fellow. This proved to be unnecessary as his late adversary was just commencing to flop about out of control and also went into a spin. He came out for a moment a little lower down and then started to spin again and continued doing so until he crashed just near Estaires, a short distance from where mine had fallen near Kemmel. Looking up to see what the enemy formation were doing, I was pleased to see that they had ignored the event and were flying off to the east. We proceeded to shoot up some of their balloons near Neuve Eglise and while doing so, Nowland, one of the pilots in the other flight, left the attack and disappeared in a dive. Thinking that he may have been hit by the fire from enemy balloons, I watched to see what happened to him and saw that he had gone down to attack a two-seater that was making back for home. After a few minutes of diving and zooming by Nowland, the Hun went into a spin and crashed between Nieppe and Merville. Then another of "B" Flight's team, Rintoul, had a go at another two seater, an L.V.G., but without any apparent result.

Some minutes later I noticed a machine lower down, apparently having a look at two Pfalz scouts that we had shot down some time erlier. I pointed it out to Bo King, and indicated that he should have a go at it, while we sat over him. He nodded his head and went down to investigate, and apparently found, by its markings, that it was another Pfalz from the same squadron. Then he and Bo staged a pretty little show while we sat in the gallery. It was obvious from the start that Bo had his opponent's measure, and it was only a matter of a few minutes before this particular machine went into a spin. We all saw it crash between Neuve Eglise and Kemmel. It was a satisfactory morning's work.

The next day, 21st May I took out a similar formation with the idea of setting fire to a balloon. Although we had fired thousands of rounds in the things they just would not burn. The job of repairing them would be a gigantic one in nearly every case where we had made a determined attack, but there was little to inspire one in the act. What we wanted was a bonfire. I let Trescowthick know what I intended to do before I left. It was my intention to get over the balloon line, select a balloon upon which to feint an attack, and then when all the hate was coming up from the machine guns, field guns and flaming onions and so on, I would divert my dive onto the next balloon, and if lucky go on to the next one again, while my pals would watch my tail for me from above. I put this plan into execution as soon as we got into the right position. Passing the sign to Trescowthick, I commenced a vertical dive on to a balloon from about five thousand feet above it. To give the impression that I intended to keep it up until right down, I opened fire immediately, and almost at once eased my machine out of its dive on to the next one in line. Then the hate started. I think everything that could point upwards was hurling stuff at me, but my speed did not give them much chance of keeping the range. I put in some ten minutes emptying my guns into the second balloon from only a few feet off, while all the time it was being hauled in. The observers had jumped out of their parachutes at my first attack and I only broke off when the thing was nearly on the ground and I could see the crew handling the winch. Still it had not caught on fire, so I fired a few bursts at the crew for luck and bolted across country to the lines before attempting to climb again. In the meantime the balloon I had first made my dummy attack upon had caught fire almost as soon as I opened fire on it, but I did not know this until after we had returned home.

Shooting down balloons may seem a simple task to which little risk attended, but it was entirely the reverse. Every German balloon had all available artillery ranged round it and machine-gun nests were placed so that they also could cover the balloon. As soon as anyone poked his nose in the direction of a balloon, everything would open up, anti-aircraft, field guns, flaming onions, machine guns, and if you wanted to get your quarry you had to go right in amongst it all. Sometimes a balloon basket would be filled with explosives and several people when they found that it was being left up and nobody seemed to care, fell into the error of thinking that the lonely looking bag of gas was easy money, and they went right in and blew themselves up.

We used a special incendiary bullet against balloons, but even so the gas was so hard to ignite that it took a lot of shooting before they would go up. The job was always looked upon as being particularly dangerous and on one occasion two of our experienced chaps collided during an attack and were lost.

The next few days brought a couple of victories to other members of "A" Flight. I had long before discovered that one or two successful flights gave a tremendous amount of confidence to young pilots and that a bold and aggressive bearing in the air was a great asset. The mere fact of being forcefully offensive when pressing an attack always gave the other fellow the idea that you were pretty good at the game, and in fact one soon developed this attitude as a permanent quality. From personal experience I well knew that a nervous or timid disposition was a big handicap. I therefore arranged in the many talks that we had on the ground, or over a couple of drinks, that one of the new members of the flight would give battle to any stray two-seaters we might come across, while the rest of us would sit up above and see that he was not interfered with by any marauding scouts. For some days these methods were responsible for the odd victories that came our way, and the victors were able to chuck a bit of a chest, which all helped to maintain the excellent name that the squadron had as a fighting unit.

Then Tunbridge went to Home Establishment and my long deferred promotion to command of the flight came along on the 25th when I was exalted to the select band of captains and flight commanders. I had destroyed some nineteen aircraft and four balloons and had not lost a pilot over the lines, nor had one been wounded while I was leading. One or two had had their machines or engines shot up, but had managed to get across to our side. I was particularly proud of this, although there was a lot of luck in it of course.

Again we settled down to the daily task of looking for the enemy; twice a day we would work up and down behind the lines, endeavoring to get into touch but without avail. Regularly we would drop our bombs on selected areas and shoot up enemy trenches, sometimes from just a few feet off the ground. This was unpleasant work as we usually got shot about for the annoyance we caused, but it kept us busy and also gave the Bosche something to think about.

The 30th May was a good day. I was to lead a squadron show on the usual job, but Malley and I had arranged to leave the formation in the hands of our deputy leaders and to do a little private hunting. Talking over the business of not being able to bring the Hun to battle, we had decided that our formations were too large, that they were easily picked up owing to the compact manner in which we always flew, and therefore easy to avoid. We thought it about time that different tactics were employed. We worked on the theory that if the enemy avoided our formations, and we arranged a timed programme so that our people would be over certain places at definite times, then we would have a good chance to find something if we kept as far away as possible. The details were arranged accordingly, and after leaving the 'drome, the two of us pushed off by ourselves. It was a very early show, 3.24 a.m. and the hour was one of the best for picking up titbits. Giving the formation enough time to be noticed over the line, Malley and I set off in the opposite direction, taking Gillis-Watson with us.

We went well over the German side, and then worked back and soon met with results. Two Albatross scouts were patrolling at some twelve thousand feet and were not expecting anything harmful to arrive from the east. We were up-sun too, and probably the first thing they knew of our presence was the pop-pop-pop-pop of our guns, almost in their ears. Both Watson and I were successful with our opening bursts, and each of the enemy scouts fell just behind the lines near Estaires and later on in the day were confirmed by our infantry as having crashed. Seeing them going down we then turned on the balloon line and attacked three in succession. After several bursts the one I had concentrated on behind Estaires went up in flames, and Malley's victim, the next one south, went up in another couple of minutes. I chased the next two down, but as I could not set them on fire we formed up again and bombed the enemy out of bed around Merville. The main patrol returned to the rendezvous in due course and we accompanied them home. As usual, although they had seen our balloons go up in flames, they had not noticed the other combats. The success of the new scheme made us eager to try it again so the next day we tried the same tactics, but without meeting any of the enemy. We were still convinced that the plan was sound. The only danger was that we might get trapped a long way over Hunland by a large body of the enemy, but we were all pretty long in the tooth by now and would watch the sky to the best of our ability and avoid anything that looked at all ominous. If caught, we had arranged to spin down to the ground and cut for home. No other aircraft could "split-s" like a Camel near the ground.

The flight had bad luck in losing Pierce late in the month. Returning to the aerodrome after a patrol, he was

The official photographer at work taking the view over the Vickers! Cobby's comment on this candid shot was "Twig the attitude of the A.F.C. artist" He is seen here on the lower wing, legs crossed and with his finger up to his mouth.

stunting over some labor battalion troops repairing the road behind the aerodrome, and as it was foggy he missed sighting a tree in his path, and smacked right into it. He had only joined the previous month. We seemed to be doing a lot of work training new people at this time for during May three new arrivals hung their gear in "A" Flight. A.H. Lockly (A.F.C.), C.S. Scobie and R.F. McRae, a Canadian with the R.A.F., being the new members. McRae turned out to be a particularly good chap on the job.

It was near the end of May that the G.O.C. Army, General Plumer, visited us. He inspected the squadron and addressed the officers, apologising with frequent grunts for the fact that the Distinguished Flying Cross, which a number of us had gained, was not available to hang upon our chests. He was particularly interested in "X", one of our non-flying officers, who wore the King's and Queen's South African ribbons, until he questioned him as to the unit he served with in South Africa. It appeared that the officer concerned had been a sergeant somewhere under the general's own command, and when he stated that he had been a sergeant and later reduced to a trooper by the General himself, the reason was asked. When informed that it was for riding along the lines in voluminous underwear, obviously at some time the property of a Dutch *Vrau*, the General gave another "Howk" and moved off. It was the only time we saw him.

On 1st June we had a morning show with eleven of our Camels with instructions to bomb St. Maur. After dropping our bombs, Watson and I left the formation and climbed over the Hun balloons, I dived on one of them and fired at close range as it was being pulled down. At about 2,000 feet,m it burst into flames and fell to the ground. While climbing to get back to Watson, I saw an Albatross following a strange Camel towards St. Venant. I fired to distract his attention and it turned towards Estaires. I turned into the sun and started to climb, Watson moving above me. In the meantime, the enemy scout had changed his direction towards Merville and was below me and directly in line with my machine and the sun. I pulled the stick back and kicked on "hard boot" and dived flat out towards him, opening fire when at some twenty yards range. Zooming up I half rolled over him and came down again a little distance behind his tail and continued firing bursts into him. The Hun machine went badly out of control and the pilot must have been seriously wounded for he tried to regain control several times. He presently went into a sideslip and his port wing collapsed. Watson then came down and we chased three more balloons down, but could not set them on fire. Watson must have shot one of the observers while he was going down on his parachute, and severed the cord as he parted company with his chute. It wasn't pleasant to see him finish the last couple of thousand feet by himself.

One offensive patrol on 12th June from Clairmarais proved to be unusual. There were two flights of us, and as the commander of the other flight was fresh from a training squadron in England and did not know much about war flying and tactics, the C.O. gave instructions that my flight was to fly above him and keep an eye on him. It was a most foolish appointment to put a novice in charge of a fighting unit of any sort, especially one that called for the highest skill and personal knowledge of air warfare, but apparently certain domestic reasons existed for his being given the responsibility. It would have been more effective to put even the most junior pilot in the squadron in charge, rather than risk the lives of the chaps through faulty leadership and ignorance of the conditions of war flying. Still it was ordered and the best we could do was to see that he and his flight were safeguarded until such time as he gained sufficient knowledge to look after himself. He started off by crossing the lines at much too low an altitude for comfort, and then began wandering aimlessly in the general direction of Russia so that in a short time we were meeting whole hordes of enemy aircraft. I had never seen so many machines of the enemy in the sky at the same time up till then and to this day I do not know why we were not all shot down, but the most amazing and inexplicable aspect of the whole business was that they did not offer to attack us. The leader of our lower flight obviously did not recognise them as enemy machines, as he would fly towards odd formations which were scattered about, go quite close to them and then fly on, and I had to follow him literally shaking at the knees. When the opportunity permitted I did get alongside of him and wave him back towards the lines, but either from supreme ignorance or egotism, he refused to do so. There was nothing for it but to stay above and bluff it out, but I was quaking. There were only five of us in each flight and there must have been sixty or seventy Bosche machines within a mile of us. I was afraid someone would become restive and precipitate trouble, and if it did start we did not have Buckley's chance. It finally fell to my lot to open fire, but fortunately nothing of any consequence developed — why I do not know.

We were moving along at about 16,000 feet when sixteen or seventeen of the enemy who were a few hundred feet higher than we were, and about a quarter of a mile ahead, put their noses down slightly and made directly at us. When they got to about eighty or one hundred yards away, I pulled up slightly until my guns were dead on the leader, and as they all came on pressed both "tits" — or as one should explain for the information of the curious, depressed the levers operating the Bowden wire control of my guns — and all that happened, instead of the harsh rattle of two Vickers firing 800 rounds apiece, there was just one "p'plonk!!" Hastily reloading with one hand as I let the nose of the machine drop again, we passed under the enemy formation just a few feet below them. To my surprise they continued on their way without any further interest in us, and I watched them join up with another large number of enemy machines further back. They were Albatross D. VIIs, mostly with an odd Pfalz or two amongst them, but I cannot to this day imagine why they did not hop into us. They would have had it pretty well all their own way if they had, as everything in sight was enemy with the exception of our two flights. Bo moved up alongside me and made some very expressive gestures with his hands and fingers that collectively derided our late "visitors", drew attention to the various enemy formations in the sky, expressed amazement at the actions of the leader of the lower flight and at the same time indicated that we would be healthier near the line. It is surprising what can be expressed in dumb language after a few months of restriction to signs. I nodded agreement, but before doing anything further, pointed my guns into Hunland and tried them again. Same thing again, only a "plonk!!" and nothing more. In case it was a matter of cold guns and stiff lubrication, I had several more tries, but nothing better than one shot out of each gun resulted. I then dived down after the lower formation, but before I got to them the leader upended and went into a dive earth-

"The habits and manner of living of those earthbound troops in the quiet periods was a closed book as far as I was concerned."

wards, followed by one of his companions. We watched them go down, but could see nothing that might cause the sudden dive, so the remaining three joined up with my flight. We lost sight of the "divers" some thousands of feet lower, but as we could do nothing about it, we moved over towards the line and waited for them to reappear. There was no machine lower than they were themselves when last seen, and there were no balloons up, so thinking they had probably re-arranged something as a method of "showing their feelings" we presently made off home.

A little later in came the leader minus his follower. He informed us that he had gone down to have a look at what he thought was a balloon, and having found out that it wasn't had made for home, but was attacked by a number of the enemy on the way and did not know what had happened to the other pilot — Jerry Martin. He said that he had arranged before leaving that if he signalled by hand that he was going down, only one machine was to follow, and that one was poor old Jerry. We got word an hour or so later from the front line to the effect that Jerry had crashed in No Man's Land in front of Meteren, and that our people in the line did not know anything more as they could not get out to him. We were all very angry at the manner in which he had gone. There was grave doubt as to whether he would be alive or not, as the Bosche troops in the trenches had kept up a hot machine gun fire on the

"... and found the geography of the land entirely different to what it appeared to be from the air."

"The country was so smashed about that it was impossible to identify anything on the ground." These three pictures are not Cobby's but have been included to emphasise the devastation.

machine after it had struck, and as far as they could see the pilot had not left the machine. The information, however, was so indefinite and sketchy that the C.O. and I resolved to go out by road to see if anything more could be discovered. After changing into our visiting togs, we left by car for the front line, or as close to it as we could get. I had never been right into the line and was thoroughly ignorant as to how things were conducted. I knew what the country in and around the trenches looked like from the air, and actually was familiar with every turn and twist of whole sectors from this aspect, but the habits and manner of living of those earthbound troops in the quiet periods, was a closed book as far as I was concerned.

We drove forward towards the line through Hazebrouck to Strazelle, the headquarters of the battalion that had been in touch with us and found the geography of the land entirely different to what it appeared to be from the air. Soon after leaving Strazelle a large notice on the roadside warned us that no wheeled traffic could proceed past that point, so leaving the car in the care of our driver, we wandered along in the pleasant sunshine. Everything was marvellously quiet and there was nothing to indicate that there was a war on, or that somewhere around us troops were hidden from sight. The quietness was in itself unnerving, and the entire absence of life or action of any sort gave us a sense of unreality. A little further back we had passed a horse that had been fairly recently killed, and about a half mile ahead we could see the remains of a village being shelled, but we could not hear a sound other than the distant "crump" of the shells. We resolved not to go past this village which we thought must be one of the forward little hamlets on our side of the line being shelled by the Hun. The country was so smashed about that it was impossible to identify anything on the ground, so we blithely promenaded along the centre of the road.

Suddenly there was a series of terrific explosions just alongside us somewhere and I made one spasmodic leap for the ditch along the side of the road without concerning myself with the cause and found that I had been beaten by feet by the C.O. We crouched there shivering for some time before plucking up enough courage to peep out and see what had happened. To our surprise everything seemed the same except for a little smoke and the deathly silence again hung over everything. We came to the conclusion that it must have been a battery of our eighteen pounders somewhere handy, but we couldn't see them and felt that it would be dangerous to try and find them, as one might walk in front of the guns as they went off. We were as frightened as though we had been in the middle of a barrage, so we decided to stick to the road. My knees were still shaking as we moved on. Presently we came on another sign that ordered us to take to the ditches on either side of the road from there on, so in we went even more astounded at the entire absence of the rattle of rifle fire or any other sound that we had always associated with the front line. We couldn't understand why a few enemy troops could not have walked right through if they wished. There seemed to be nothing at all to stop them.

We could still see the village ahead that the Bosche was slinging some "hate" into every now and then. It seemed to be all a drab brown and was practically flattened out and with the dust and smoke was unrecognisable. Anyhow we only had a very vague idea as to where we were. After a while even the ditches flattened out and although we were surrounded by piles of bricks and debris, the formation of the road was still discernible so we went on. We expected at any moment now to come upon the front line trenches bristling with troops and to find our way barred by wire, but we did not realize our own ignorance. We had just passed a pile of bricks and were about to move out on to a piece of fairly flat ground to have a decent look at the country in front, when we startled by a voice that appeared to come from the middle of the pile of bricks just behind us.

"Get under cover you pair of so and so fools", the voice growled. "If your life is not very valuable, mine .... well is. If you stay there more than a second, you will be sniped with anything from a minnie to a six inch shell." It seemed to us rather stupid that whoever owned the voice should be so peevish in all this quietness and with the war so far away, but we wanted information and as this was the first chance we had of getting any, we obliged by dropping into a trench, which the voice had directed us to as being "for damned fools like us."

The C.O. then asked if "the voice" had any knowledge of a Camel coming down on the line somewhere in front of Meteren, but the terseness of the reply made it quite clear that the owner of the voice did not know anything about the matter, and what was more, didn't want to.

"Have you any damned idea where you are?" he asked. The C.O. replied that we had only a vague notion, but guessed that we were up near the line, somewhere in front of Meteren. "Too damned right you're near the line" came back, "Do you know how near Meteren is to the line?"

"Yes, it's about three hundred and fifty yards on the Hun side of the Bailleul-Cassel Road" I replied. "Well, just have a look about you", snapped back our peevish friend. "That village that's getting hell pasted out of it just in front, is Meteren, and we're about one hundred and fifty yards from it, and they are my guns that are doing the job."

We were in front of the line! I sought information as to how we could get back and the voice directed us to follow the trench we were in, and we would find that it would swing round into communication trenches further along, and we could pick up the road again about a mile back. It was also suggested that we keep our heads down as anyone — friend or enemy — would have a pot at us if they saw us.

About an hour later we got onto our road again and I was never more thankful to see a bit of clean open country, but we never found out anything more about Jerry Martin.

When we got back home to Clairmarais, I found out what had been wrong with my guns. They didn't have the muzzle cap attachments on and apparently it was my own fault more or less. These attachments are for collecting the gases of the explosion and are to help push back the recoiling portions of the gun, so that it automatically picks up the next round each time the gun is fired and reloads it. The trip before this one, I had noticed that the guns were much slower than usual and had complained to the gunnery officer about them. Then having a few minutes on hand, I had had a look at them myself and found the muzzle caps were pretty dirty with carbon. As they had to be polished brightly to be effective, I had then cleaned them up and put them back. The armorer coming along later had examined the guns and had seen that the caps were clean, and as I had particularly mentioned them in my complaint, he had left them on the wingtip in the hangar for me to look at. Then when my bus was wheeled out from the hangar a they had slipped of unnoticed and I had taken off without them. It certainly taught me a lesson.

On the evening of 17th June I led fifteen Camels out from Clairmarais on an offensive and bombing patrol. The objective for our bombs was again Bac St. Maur, and after letting them go at the village, we moved off up the line. Watson and I had again arranged to leave the patrol if no enemy aircraft were engaged, so we pushed off by ourselves. After about an hour's flying up and down the line at 4,000 feet, I observed five enemy scouts in an extended "V" formation about a thousand feet lower than we were and to the east. Moving along the lines to get the sun behind us, we maintained a position relative to theirs' except we kept the odd thousand feet up our sleeves. After some time they turned to the east to go home and Watson and I immediately dived on them. Their formation left them very vulnerable to attack from the rear. The leader was an Albatross and he was flying at about 4,000 feet with the next two machines about six hundred feet behind and above him. The last two machines were in and about the same relative positions to these two. We were thus able to drop on the two rear out of the sun, without the other three being able to interfere until we had a few bursts at them. Further, we would be able to pull out if they should see us coming before we got down, without any risk to ourselves. Watson took the rear left man and I the right one, and we both held our fire until right on top of them. I think that until we opened fire, they were entirely ignorant of our presence. My first burst accounted for my fellow; his lower right hand plane collapsed and the machine went into a spin. Both wings came off lower down and the machine crashed east of Laventie. Watson's victim had already burst into flames. The other three enemy scouts had now turned to protect themselves, but my initial speed gained during the dive, enabled me to do a high Immelmann turn and I had time to pick my direction of attack on a second machine. I fired several bursts at this one, and a general dogfight ensued. I stuck on the tail of my opponent, firing short bursts, and whilst trying to tighten up the turn to get my guns ahead of him, I was nearly startled out of the cockpit by a pop-pop-pop almost in my left ear. There was a bang somewhere alongside my face and a smack across my left cheekbone. I yanked the machine up as high as I could and looked over the side, and there was Watson hanging on his prop, firing at one of the other machines beyond me. I had apparently collected some of his fire. In the meantime my opponent had gone out of control, obviously badly wounded or with his machine badly damaged. He tried to flatten out near the ground and land, but in doing so struck a tree and crashed in a field. I went down and dived three times at the machine and fired about four hundred rounds into it, but the pilot did not move after crashing. Watson had crashed his second victim while I was following mine down to the ground and I saw it spin into the ground three hundred yards or so from where my first one had fallen. I had a few words to say to Watson when he got home, about looking where he was pointing his guns, and showed him the cut in my cheek. Apparently one of his phosphorous bullets had struck my left centre section strut and had split it, a piece cutting my cheek and going out through the glass of my goggles. It was only a scratch, but gave a good deal of trouble as the phosphorous hotted it up a bit and it would not heal. For many days one side of my face looked like an apple dumpling.

On the next day, the 19th, I was approaching the line at 7,000 feet with a formation of seventeen Camels and saw four Pfalz scouts trying to cross the line with what was obviously the intention of attacking our balloons. Three of them turned back on sighting us, but one continued on and dived on the balloon at Hazebrouck. I dived east to cut him off when he saw me and tried to get away, but I got on his tail and put about a hundred rounds into him from point-blank range and he crashed into the forest of

A double aerobatic feat. Two Camels after a taxying accident at Clairmarais.

Nieppe, on our side of the lines. We discovered afterwards that there was something extraordinary in the manner of his clothing. He was dressed in a swallow-tailed Beaufort coat, very much worn, breeches in a very bad state of repair and very dirty, and was without collar or tie. He had not shaved for some days and looked as though he had come off a "bender". We never discovered the significance of his dress, but he was vey much shot about.

A number of full strength patrols were staged over the next few days, our objectives being the enemy on the ground. The air was quiet and the German airmen were keeping well out of sight. On the 25th, during one of these patrols, we saw three Pfalz scouts below at somewhere around 9,000 feet, between Estaires and Laventie. Watson and I dived on them, and I chose the rear one, firing forty rounds or so from close up. The machine went into a vertical dive for some distance, then came out and went into a spin. The top plane came off and the wreckage fell to the ground. Watson had several goes at his but the Hun eluded him and got away. We then dived on an A.G.O., a two-seater about a mile away, but I overshot and on turning saw Watson close in on his tail. His tracer ammunition seemed to be going through the machine and presently I saw it turn on its back and go down out of control and crash to the south east of Estaires.

Our new tactics were certainly bringing home the bacon. Almost daily we were concerned in fights, and although no big dogfights occurred, we felt on top of things. It was our ambition as a squadron to be regarded as second to none, and we seemed well on the way to the coveted position at the top of the victory list. We had a great deal of self confidence by this time, so much so that we regarded odds of two-to-one against us as being about an even contest. This was particularly so when those of us who had been together for so long, such as Malley, King, Watson, Trescowthick and one or two others, went out together. We had so much knowledge of each others' tactics and such perfect understanding, that we would have tackled almost any number of enemy aircraft. Our arrangements for engaging any smaller number of machines than our own when on patrol with a big show, were very simple. If we could dive on them, the leader, who was usually myself, would signal that he was going down. The other flight leaders would come down with me and our deputy leaders would take over and stay above us, or in such a position that they could keep venturesome enemy scouts, if there were any about, from coming down on top of us. This left us free from worry as to what might drop out of the blue. If the presence of the whole formation was required to participate in the business, then the leader would signify his intentions by rocking his machine backwards and forwards, which meant "do as I do." No Camel was allowed to leave the formation under these circumstances until battle had been joined.

June 26th was another day in which we applied these measures successfully. Trescowthick had I had decided on a little independent war of our own and leaving the ground at 3.15 a.m., had gone over in the dark to call on some aerodromes and leave some nickel-covered visiting cards. After dropping some fifty pound incendiary bombs on some hangars near Wavrin, we had shot up the quarters and then emptied our guns through the roofs of both the quarters and the hangars. We hoped that we made ourselves felt around the place and then scooted back for the lines. We were out again at five past seven after an early breakfast, three flights of us this time, and on this show, Watson crashed another two-seater, an A.G.O., near our favorite cemetery, Estaires. A Pfalz scout was unable to put up much of a fight over La Ventie a few minutes later, and also went out of control. He was alone on the line. In the evening I was again out leading thirteen of the chaps on patrol to look for a scrap and from over Armentières at 15,000 feet, saw a patrol of five Pfalz scouts some two thousand feet below in the direction of Lille. King, Watson and I dived on them. King took on the top one, Watson went on to the rear one, and I went after two that endeavored to get back east by diving. I cut them off and engaged the rear one, and after manoeuvring for a few moments, managed to get on its tail. A series of short bursts from my guns and it turned over and plunged out of control with smoke pouring out from behind for over 5,000 feet. There was an evening ground mist covering the country, so I was unable to see it crash. Both King and Watson shot their victims down in flames and all three fell between Armentieres and Habourdin. I could not see any sign of the other two, so rejoined the formation that was till keeping guard above us. It was a long but effective day and we felt entitled to quaff an odd glass or two of the cup that cheers, although our spirits were high anyway.

Malley, another pilot named Crosse, and I went off the next day on what we called a special mission. We did not always disclose what we were going to do on these occasions, in fact we did not always know ourselves, but relied on something turning up. If things were quiet we might only give the troops on our side some stunting to cheer them up, the main idea being that we worked off surplus exuberance. On this trip, however, we went over to an enemy drome on the outskirts of Lille and "left cards". There were two Albatross scouts on the tarmac, both painted black and white so we commenced by setting them on fire. We then chased everybody away that put their noses into sight, and finished up by flying across the aerodrome with our wheels on the ground and firing into the open doors of the hangars, hoping that we would burn up a few more machines, but we were not successful in this. We then crossed Lille at some twenty feet above the roof-tops, waving our hands to the people in the streets, and quite a number of people waved back. We were fired at by the forts around the town, but we kept right down on the ground all the way back, twisting this way and that so that enterprising field gunners and people with machine guns could not get much of a shot at us. As a matter of fact we were pretty safe, as they would not expect the enemy to come coursing across country from behind them, especially so low down.

It was amusing to see odd batches of troops come out of billets and hutments and start to wave to us, and then go to ground like startled rabbits when they realised that our machines belonging to the enemy. We may have hit some of them with our wheels here and there. We usually zoomed up after passing them and looked down over our tails as they lay on the ground, and facetiously blew kisses to them. They thought us quite mad I expect, but it must have had a demoralizing effect on them to see us sporting round their territory without hindrance.

We had a laugh over the episode that night in the mess, and Bo King wanted to know whether the girls in Lille were any better looking than the ones on our side, but we could not enlighten him. He made us promise to include him in the next joyride. He reckoned that he wanted to drop a bouquet on the best looker he could find in one of the big towns behind the lines, but we thought we would defer it for a while as the Bosche might be waiting for us if we repeated it too soon. But Bo's suggestion had given me an idea that I intended to put over as soon as the opportunity offered.

We assisted the 5th British Division in one of its stunts on 28th June by indulging in a trench strafe of the enemy in front of Nieppe. Trench strafes during any sort of show were not pleasant to participate in. They always required one's presence practically in the enemy trenches, and of course our part was to keep the enemy down or do as much damage to him as we could. Quite simple!!! But of course our artillery, both large and small, would paste hell out of the particular piece of country under consideration, which might stretch up to several miles. As soon as our shells began to fall, the enemy would open up on our trenches and try and knock out our artillery as well, and if one had the power to see the shells as they passed, a view down the line would look like a gigantic elongated fountain, only instead of water there would be shells from small stuff up to sixty pounders. The greatest concentration, of course, would be in the vicinity of the trenches, and with shells falling from both directions every yard or so. In amongst all the hate we had to give an outward semblance of enjoying ourselves by diving practically into the trenches and contributing our quote of annoyance, while they returned the compliment.

It was very nerve racking and it seemed to fall to us to do that sort of thing once or twice a week in normal times, and several times a day on important occasions. I know that I always emerged from these shows almost rigid with tension.

On this day, the 5th British Division had assaulted the enemy's lines at dawn and driven him out of his positions

When a squadron moved, it was a major upheaval. For a summary of squadron equipment and methods of transport, see the Appendix on Page 105 dealing with squadron organisation.

near Vieux Berquin. At the same time the 1st Australian Division carried out a most independent raid in broad daylight, and took prisoner almost the entire German formation and their trenches in front of Merris. It was a most cheerful sight to see and it looked from the air as though they just walked over and said "Thanks, we'll have this."

We kept an eye on this area all day and apparently the operation was entirely successful. Late in the afternoon Bo and I went to try and intercept German aircraft reported to be flying in the vicinity. We ran into twelve machines scattered around the sky in small formations, and selected five two-seaters making towards Vieux-Berquin. I was fortunate in getting one almost immediately on diving at it, and it zoomed up as the pilot was hit, so I had to continue my dive under it to avoid a collision. On recovering I looked back and saw a number of enemy scouts converging on me from several directions and from above. My speed was very high, so I zoomed up towards the most favorably placed one and when almost on my back, below and in front of it, I let go a long burst. I did not stay to see the result, but made for the ground and thence across the line as quickly as I could. The other machines cleared out.

St. Omer was one of our regular ports of call in our moments of leisure. About once a week we used to put on our "posh" uniforms and push off by tender to the Officers' Rest Club, bathe, feed the inner man, drink deeply of the wine of the country, and then look the place over. After our odd jobs of shopping had been completed, we would rendezvous at the club and carry on the work of downing copious draughts. One of these occasions became a pretty full day and will illustrate the opposing point of view of the chaps fighting in the air with that of the unfortunate fellow in the trenches. It may also serve as an example to the young, of the awful effect of gazing too long upon the wine when it is ruddy, or any other color for that matter.

Two officers from a Scottish regiment were in the club on one of the occasions, on their way back to Blighty and leave. They had apparently "lorry hopped" straight out of the line and were covered in mud from their khaki aprons down, and offered in their war-stained equipment, a silent contrast to ourselves in our spic and span uniforms. The comparison was not limited though to our opposed appearances. Although admitting that the man in the air sometimes did a job of work, they were far from silent upon the ease of living that seemed to be ours in comparison with the discomfort and entire lack of amenities in the trenches. As we alternately stood the drinks (there were eight of us) they became more and more insistent that we had "a damned easy time" of it and looked more like a nice lot of little schoolboys than "fechtin" men. The mellowing effect of the potions we were taking, combined with a growing strength of their argument, reacted in a peculiar way on one of our number. Instead of either laughing at them, which would have been in keeping with his cheery nature, or bumping their heads together, which he could easily have done with his six feet of brawn and bone, he took it to heart. For a while he sat mulling things over in his mind. At last he decided that if we weren't going to earn our keep, he was going on his own. More in the hopes of persuading him against spoiling the day, and night, a couple of us went with him, but found that his ire had been aroused. More, he was in a peculiarly sentimental state of mind and was a little bit mixed as to whether he was going to take the air because of the lighthearted insults that had been passed, or because in his cheerfully mellow state he honestly thought we were not doing enough, or whether on the other hand he wanted to avenge the muddy and bedraggled appearance of our erstwhile Scot acquaintances. Whatever the urge, it was sufficient to make him adamant in his intentions, and as we were quite open-minded and inconsequential about the whole thing and regarded it as a bit of a joke, off we went on the forty mile run back to Clairmarais, wheeled out our buses and took the air. We were determined in our condition of exhiliaration and exuberance, to clear the sky. But alas for the plans of mere mortals, the oft repeated "man proposes" was never more aptly applied than to our actions for the next couple of hours. We cruised miles over the line, chasing around enemy aerodromes and did everything we could to get into trouble, but one would imagine that an armistice had been declared as not a single enemy machine of any sort was sighted. The last amusing effort of the one who was the cause of it all was to fly upside down quite low down, and fire colored Verey lights into a village on the enemy side that had turned out to view the peculiar circumstance of a British pilot devoting a nice bright afternoon to giving an aerobatic display to the enemy. Even after the first few moments, the clear fresh air had cleared the haze from our heads, and I for one was quite fed up early in the proceedings and indicated to my companions as soon as possible, that the game was off. We returned to St. Omer with our chests out, full of satisfaction (more or less) looking for our rugged Scots, but when we got back to the club neither those kilted jokers nor the other members of the party were to be seen. Usually things petered out in somewhat the same way; we would become a little mellow as the time advanced, there would be a round of choruses at the piano and the inevitable cheerful but tiresome wallah who wanted to sing something and was still wanting to sing it when we were wanting to go home. Others would gossip on their own particular phase of the war and what they would have done if they had been in charge of affairs, while others were quite definite that they would have done something entirely different from what Haig had done on the Somme and so on. Flying people usually had a corner of their own and waged bloody warfare in the skies, amply illustrated by beer dipped fingers, drawing on the polished table tops. Many an air fight was brought to a successful conclusion during these sessions. Someone would start off with an anecdote, and his neighbor would cap it. "I was at about fifteen thousand feet", he would interject, marking himself in space on the table with a finger that he had immersed in his drink, "and the Huns were about five thousand feet above me." More marks followed to indicate Huns. "All of a sudden they came at me hell for leather". (pause while another draught was taken). "I splitarsed like blazes and did an Immelmann!" Momentarily the speaker would abandon his clutch on your arm and remove his finger from his beer or whatever it was he was using as a medium for draughtmanship, and seize an imaginary control stick in his hands, pull it back into his side, kick on a hefty amount of boot, and then pause while you assimilated his manoeuvre. "Then pressing both tits, I dived right into the internals of a Pfalz that got in my way," he would continue. "Yes, the poor devil went up in smoke and never even knew what hit him."

An SE5A of No. 74 Squadron R.A.F. after a landing accident at Clairmarais.

We would all nod understandingly while the narrator would glower at us over the remainder of his drink. We were all heroes on these occasions.

Bo King and I thought that we would have a look-see the next day about 2.30 and proceeded over the lines towards Outtersteene. We had been warned by the wireless people that they were picking up enemy signals from a German artillery machine from this direction and on getting round the bank of cloud I saw an L.V.G. making for Merris, about a hundred feet below me. I dived on this machine and fired a burst from close in behind but without any visible effects. I then put my nose down for a moment or two and zoomed up vertically under the pilot's cockpit, firing when right up close. The machine swayed sideways into a steep sideslip and piled up just outside of Outtersteene. A few seconds later several scouts came down on me from behind the clouds but after firing a few shots in my direction they all pushed off with the exception of one Pfalz, which was turning up towards me from about 600 feet below and firing as he came. I flicked my bus over into a half roll so quickly that he lay directly under me as I came out. Hastily pressed the triggers, I fired off a long burst at him. He started to spin and almost immediately burst into flames. Malley, out on a "pleasure" cruise, had seen the scrap from some distance away as he was coming up, but it was all over before he could intervene. In the meantime Bo was missing and we put in some time looking around without seeing a trace of him. I was afraid that he had been caught by some of the Pfalz scouts that had dropped on to me but we flew all around below the clouds to see if there was any sign of him on the ground. We could see nothing. I returned home feeling pretty blue and hoping that he had turned up, but he had not arrived.

Malley said that he had not seen a sign of him. He had seen the five scouts dive on me from a long way off and started to climb to the rescue. Although he had me in view the whole time, mine was the only Camel in sight. We were still waiting on the tarmac twenty minutes later when we heard a Camel coming. Presently Bo came racing down the drome low down and waved to us as he went over our heads. We heaved a great sight of relief. When he came in he told us that he saw the Pfalz have a go at me and had "pigged" in on the tails of them from behind a cloud, and had followed them well back into Hunland but without visible result. He had also put in some time searching for me but had given up. Now here he was, and what about a drink? It was duly taken.

Once more just before 6.30 Bo and I went off together to see what things were like in the air. We crossed the lines at Merris and saw a number of small groups of enemy aircraft over Hunland. There were three two-seaters making towards Vieux-Berquin at some four or five thousand feet lower than us. We immediately attacked from behind the clouds, and were on them before they could avoid us. I fired a burst at the nearer machine without effect and as King was also engaging it, I dived under it and came up under the tail of a second one. I fired about two hundred rounds at him from close range before he fell into a vertical dive. He then turned onto his back and finished by spinning in this attitude. It maintained the spin until it crashed near the road north west of Messines. I saw that King had accounted for his opponent by this time and it was also spinning earthwards to crash about half a mile from my victim. Bo's machine was an LV.G., and mine a Halberstadt.

Having some time left before it was necessary to return home, I climbed into the clouds again to see if I could find anything of further interest. While over the river Lys, I saw a balloon in a very tempting position and dived on it from 4,000 feet in a vertical dive. I was using special ammunition for firing balloons. The big sausage started to smoke within a few seconds while it was being drawn down,

Pilots of No. 4 Squadron's "C" Flight led by Captain E.J. McClaughry (centre, over the bomb). To his left were Lieutenants J.H. Weingarth and R.H. Youdale. The others are unknown. At this time the squadron's aircraft were identified by a white vertical bar in front of the fuselage roundel in addition to their individual letters.

and when nearly on to the ground it burst into flames. Bo King was waiting for me when I came up again. He had been watching my effort while doing a little private bombing of his own.

We had a taste of what our bombers were doing to the enemy over the line while on this 'drome, but its intensity and effect was far greater than anything we had ever pulled off. During the moonlight nights around the Whitsun period, the Bosche descended upon us with droves of bombers. Of course our night bombing aircraft may have been doing just as well on the German side, but to us in fighting squadrons it savored of unfair tactics. Wave after wave of night flying machines would come in over the salient and play merry hell over the triangle of Ypres-Haazebrouck-Omer, with others going as far as Etaples, Boulogne and even England. To us it was one of the most nerve-racking experiences of the war. All the roads, concentrations, dumps and villages received attention and with pretty big stuff too. Shell dumps were blown up, some were burning for days and days and for a couple of weeks it was like living in an inferno. Not that we had many casualties from the attacks, but what with the continuous drone of multi and twin-engined machines, which sounded so much like (Where are y'r dumps — where are y'r dumps) the bursting of their bombs and the explosion of the dumps going up, combined with the banging away all around us of our seemingly innocuous anti-aircraft batteries, and the noise of their "duds" and nose pieces whistling around us, sleep was out of the question. Added to this was the excitement of watching our searchlights picking the raiders up in the beam and our night-flying fighters giving them a bit of hurry up. The "Archies" would open up and as usual get much closer to our fellows than to the enemy. Then one would hear another flight of bombers coming our way and listen to the line of approach and try to estimate whether we were in their tracks. Then would come "crump" after "crump" as they let go at targets below.

One evening about twelve o'clock, two big bombers came right over the 'drome at about 1000 feet and had a smack at us. We could hear the bombs whistling down on us and everyone threw themselves flat on the ground. We heard the first two hits but there was no explosion. Then the other machine had a go and there was a tremendous bang from the centre of the aerodrome followed by another a little further off. Next morning we looked to see the result. There was a hole in the aerodrome that would take about four double-decker omnibuses where the first bomb had exploded, but our luck was remarkable for the first two had hit in spaces between hangars and had not gone off. The holes bored in the ground by the bombs were about ten inches across. Later, when engineers had dug them out for us, we realized how fortunate we had been. Each bomb was about 850 lbs apiece and stood nearly as high as a man. If the German armorer responsible for fusing the bombs knew that after getting in such excellent shots they had failed to detonate, I am sure he would have felt like kicking himself to death. He had left the nose vanes locked up when the bomb struck and cousequently the striker pin had not been rotated up to its position for touching off the explosive. This vane, which acts like a windmill, is locked for safety until the bomb is on the machine. It is connected to a thin rod, and when the bomb is released the rotation of the vane screws the rod up against the very delicate explosive material that does the initial work of sending off the bomb. The last task of an armorer before a machine leaves the ground is to free the vane and drop a finger or stop on the bomb rack between the wings of the vane to hold it still in flight. There would be many *Mein Gotts* and *Donner und Blitzens* when the pilot, who would have been chivvied around the sky a lot before he found a target for his bombs, and who must have had a thoroughly uncomfortable couple of hours, asked his mechanics on return what was wrong with the bomb department.

The intensity of these raids was so demoralizing that we were ordered to pack up camping kit and move away some miles and sleep under hedges, but several of us stayed behind, and although we did not completely get used to the nerve-racking din, a few bottles of Bass helped us to get in a fair amount of sleep.

New pilots again came along during June. J.S.M. Browne, a Canadian, came and went, nobody knowing whence. He was on high interception work soon after his arrival, well on our our side of the line and all we know is that up he went on his job. To this day no word has been heard of him. If he had been shot down he should have fallen well on our side, but if he came down either alive or dead on the German side, the enemy would have reported it. They were very meticulous on this. The only conclusion to be drawn is that somehow he fell into the sea, but it is difficult to see how this could have happened. There was not a cloud of any description in the sky. It was perfectly clear for a hundred miles and one could see half-way up England from the line. The fact remains that he was never heard of again.

The youthful Pate, our pleasant English laddie, also piled himself up when taking off by colliding with another machine and breaking a leg. He was quite all right otherwise but we never saw him again. Our old member Roy Burton, one of the originals also arrived again and departed, and a newcomer R.T. Edols joined the "A" Flight band.

June was an unusual month in many ways. We had always maintained a very high degree of serviceability with our aircraft, but the average daily serviceability at dawn each morning, was 24.2 and this out of a total of 25. This was unusually high and reflected great credit upon the ground crew of the quadron. Their work was always excellent and their keenness remarkable. No other Camel squadron even got within two or three of our daily average. Further, with the twenty-five machines we averaged 47 hrs. 7½ min. per day in the air, and one of them, the C.O.'s machine, was not a regular performer. In addition, not one Camel left the formation for any cause, guns, aircraft, or engine for the whole month. This was all achieved in spite of the fact that the squadron suffered a severe epidemic of Spanish influenza during which nearly everyone, pilots and men, were laid up for several days. But whilst this month was outstanding in these respects, it did not exceed the normal month by very much.

Pilots of No. 4 Squadron as on strength on 10th June 1918. Front row from the left: Lieutenants V.G.M. Sheppard, H.G. Watson, Captain E.J. McClaughry, Major W.A. McClaughry, O.C., Captain A.H. Cobby, next man unidentified, and right, Lieutenant J.H. Weingarth. Immediately above Major McClaughry was Smallwood, and above Cobby was Roy King. The rest of the names are unknown.

Captain A.H. Cobby, now as Flight Commander of "A" Flight. In June 1918 No. 4 Squadron became part of the same R.A.F. wing operating out of Reclinghem and engaged in enthusiastic and friendly competition for the highest number of enemy aircraft destroyed.

Good Hunting

CHAPTER 7

We had another move at the end of June, caused by persistent attempts to bomb us out of Clairmarais. Although not the slightest damage was done to aircraft or personnel during these attacks, the strain became so great that it was considered advisable to shift us out. Our new home was Reclinghem, about twenty miles to the south, and south west of Aire.

The situation here was the most pleasant that we had been stationed in so far. Our quarters were in an orchard, but the trees had long since given up the unequal struggle and refused to come to fruition. We lived in tents for the first time, and after the dingy packing cases that we had slept in on the previous 'drome, this was most refreshing. We put up permanent structures for the messes and cookhouses and were most comfortable. Also for the first time we were on the same 'drome as another Australian Squadron, No. 2, under Major Murray Jones, M.C., D.F.C. This was a scout squadron too, and we anticipated good hunting together. Murray, I knew to be a stout chap as I had served with him in the pre-war days when we were budding young officers both in Senior Cadets and Infantry. If the rest of the chaps were anything like him in character, they would be a good team. Some of them we already knew and it was pleasant to resume old friendships.

As a matter of habit we still kept up our irritation tactics of bombing everything on the ground that looked like being worthwhile. Merville and Estaires were always good hunting and camps, dumps, billets and anything moving received plenty of attention. I would have hated to try to exist in that area. It had been most uncomfortable at Clairmarais when the Gothas and Friedrichshafens used to come over each night, even though an odd bomb or two only exploded within a mile of us. What it felt like round this particular sector could only be imagined, as we would put a hundred or two bombs into it daily.

On 2nd July I went out alone in the afternoon with the intention of having a go at any balloons that might be up, providing the sky was free of the enemy. I climbed up to 10,000 feet to have a look around before crossing the lines, and saw a Fokker Triplane flying in a north-westerly direction under a big bank of clouds over beyond La Bassee. It looked too good to be true, but a series of white

"Archie" bursts some little distance away made me suspicious of a trap, so I climbed away from the lines. It was the habit of our anti-aircraft people to put up a few bursts to warn us if there were any enemy about, and my suspicious were well founded. On the far side of the clouds, which were rather broken, was a nice little team of four Albatross scouts, just waiting for some innocent to come along and have a go at the plum waiting to be plucked, but I had played the same game before, so I played possum too. Turning away as though "not having any today thank you" I disappeared to the west. As soon as I was out of sight I pushed off south as hard as I could lick and crossed the lines lower down and made well east before circling round behind the bank of cloud under which the decoy was hovering. By keeping well up into the cloud drift I was able to sneak up close on the unsuspecting triplane, and when judging I had gone far enough, dropped out of the "murk" and found the enemy just ahead and a little below me. He was pretty wide awake as he pulled his machine around in a "split-S" turn before I could open fire, but I was in a better position, and pulling my bus around after him I got my sights just ahead of him and pressed the triggers. I was only about fifty feet above and behind him, and he heeled over out of control and fell through a lower bank of clouds. I followed him down, but the protecting aircraft from above had spotted me and were on their way to avenge their loss. I went straight into the clouds and dived towards our side, which was only a short distance away. There was

Major A. Murray Jones, O.C. No. 2 Squadron. "Murray I knew to be a stout chap as I had served with him in the pre-war days when we were budding young officers both in Senior Cadets and Infantry."

The Fokker D VII. This fine aeroplane represents the peak of German operational fighters during World War I. Its speed, manoeuvrability, strength and firepower made it a favorite with its pilots. It was often brightly decorated as on the example shown here. This example shows the thin type of cross in use in late 1918.

SE5As of No. 2 Squadron's "B" and "C" Flights lined up at Reclinghem. It is interesting to note that some aircraft had their white serial numbers painted on the fin whilst on others it was displayed on the fuselage. The wing upper surface markings are also of interest.

no sign of them when I came out so I recrossed the lines later and found that the "tripe" had crashed north-east of La Bassée.

On 7th July, Trescowthick of my flight led a patrol of Camels on escort to some D.H.s that were going out to do some bombing. A new German scout machine, the Fokker D.7, had just arrived at the front and it was superior to anything we had at any height above ten thousand feet. Fourteen of these dived on Trescowthick's formation and shot it about considerably, but they all got home. It was the first time for months that German machines had had the temerity to shoot any of "A" Flight's buses and we did not like it. It certainly looked as though their superior performance would lead them to do it again, unless we were very careful.

German artillery machines were again becoming active on our sector, and while patrolling alone along the line on the evening of the ninth, I spotted two of them to the north-east of La Bassée. They were evidently working together for mutual protection. I attacked one of them and fired a long burst from above and behind and then went lower and came up underneath its tail, putting two bursts into it from close up. After wallowing about for some time it went into a spin and fell to the ground a little to the north of Gravelin. His companion, after firing at me during my approach, cleared off to the east and did not return. I then tried to destroy a balloon nearby, but got my machine shot to pieces for my trouble. I had all my work cut out in getting home safely. Most of my controls were gone as the result of fire from the ground. All told I had forty bullet holes in different parts of the old bus.

We kept on getting reports of the excellent fighting ability of the new Fokker. We had not come in contact with it to any extent as most of our patrol work was being done at lower altitude, but our fellow Australians in No. 2 Squadron were continually meeting them. Their S.E.5s could get to greater altitudes than our Camels, which were at their best up to about twelve thousand feet. We could get much higher, of course, but the performance fell off rapidly above this level, and against the new Fokker, would put up an indifferent show. Later on we did meet them up higher and managed by sheer hard flying to hold our own, but unless one was an exceptionally good pilot the odds were definitely not good.

We dropped in at the 1st Australian Division sports at Wallan Capel after a patrol one day. We finished our job over the line as soon as we could and then went over by arrangement to give them a demonstration in the air. Things did not work out as intended though. After giving them everything we could up top we finished by dropping all sorts of appropriate messages — Trescowthick confining himself to the parachute and bed room utensil — in which he had placed many cheery notes.

We had decided to land and pay a call of half an hour or so as sports of this nature were always worth a visit. Seeing a field of suitable dimensions a few hundred yards away we made for it, and as soon as I got my tail down, I looked back to see how the other two chaps were getting on. Trescowthick was all right and well behind me, when without warning, my tail started to go up and the view ahead changed from half sky and half ground to all ground. Then I was looking at the grass just in front and next moment I was upside down with my face being pushed into the ground.

When picking out this field everything looked nice and green and level from up top, but about two thirds of the way along a wide sloping ditch some six feet deep and fifteen feet wide ran across the ground. It was well grassed and from the air looked quite level.

The centre section collapsed and there I was in the undignified situation of standing on my head and unable to extricate myself. The efforts of Trescowthick, who had just arrived, could not pull the tail over the free me. After a few moments a bright and cheery red face appeared alongside mine and amidst much puffing, the owner thereof demanded to know if I was unharmed. Upon being assured that I was, more or less, he said, "Help liquidate then", and passed in a bottle of whisky. A practical man that one, a padre of some renown by name of Dexter, well known to many. Naturally, I nearly choked trying to drink spirits out of a bottle whilst upside down, but the quickness of mind was to be admired. Taplin, the third member, did not land but flew off home as soon as he saw I was O.K. It took three days to leave that sports meeting. One for a heavy lorry and trailer to get there with a breakdown team, one to dismantle the machine and one for the team to have a rest. I think I met every man in the 1st Division during that period and even essayed to win a horse race. It was too long to wear flying boots and helmet though, and no one by arrangement would lend me shoes or hat. It was their idea of a joke that I should be known wherever I went.

Trescowthick and I destroyed a two-seater and a balloon respectively in the early hours of the fourteenth. I had gone down after the balloon and set it on fire, when an A.G.O. two-seater pursued me and opened fire at me from long distance. Trescowthick, however, was watching and hopped on to its tail while it was diving, causing it to go straight on and hit the ground without even coming out of the dive. Again on the following evening Wattie and I had a bright session with five Pfalz over Armentieres. They were flying in company South East from the line towards that much battered town at 6,000 feet and my partner and I climbed above the clouds from further east and flew towards them. The clouds were scattered cumulus, some thousand feet thick and when we had judged we had gone far enough, we put down our noses and went through them flat out, Wattie's wingtip almost brushing against mine. We came out a little to the east slightly above the Germans and immediately attacked. There was a crowd of other enemy scouts about at that time, including some Fokker triplanes, a type we had not seen for some time, but they could not have noticed us. We were almost behind the Pfalz, and after turning slightly whilst diving, came down onto the tails of the two rear ones. My first burst set my opponent on fire. Wattie's aim must have been accurate, as his man went up nose first until the machine stalled before going down in a spin. I did not watch it for long, as I went through onto the next machine on my side, opening fire at once and putting in a long burst as I came up on him. He half rolled at once so I zoomed up over him, and turned to fire again. This was unnecessary, however, as the right hand wing had collapsed and the rest of the machine was spinning down crazily. I was startled by the familiar pop-pop-pop-pop of machine guns from above, and glancing over my shoulder, saw the Triplane formation coming down on us. Wattie also shot past me on a "split-S" turn firing his guns to attrack my attention to the danger, and pointing upwards. I shook my head and pointed downwards indicating "nothing doing" then kicked my bus into a half roll, pulled the stick back into my tummy, and went into the clouds in a vertical dive and kept in it for abut four thousand feet before easing out. This manoeuvre was not as dangerous from an attack point of view as it may seem, as the "tripe" was notoriously weak structurally, and could not dive like us. I came out into the clear pointing south so I eased by bus round slightly towards our

A.H. Cobby all wrapped up and ready to go. Note the bomb racks under the fuselage. In this instance only two racks had been mounted for two 25 lb. Cooper bombs.

Captain R.C. Phillips

Some of No. 2 Squadron's SE5A exponents. As with No. 4 Squadron, some had greater aptitude for aerial combat than others. These pilots all finished the war as aces, having shot down five or more enemy aircraft. The ranks mentioned are at the time of the photograph.

Captain H.G. Forrest

Captain F.R. Smith

Lieutenant R.L. Manuel

Captain L.H. Holden

Lieutenant A.T. Cole

Lieutenant G.H. Blaxland

side, and looked back for Wattie. He had come through some distance behind me and was already pointing towards home, but as we were only some three hundred feet or so off the ground and about half a mile from the line, we stayed apart to divide the attention of the machine-gun enthusiasts in the trenches, and shot across the lines to our side. Only one Triplane came through after us but he was a good distance away and as soon as he saw that we were out of reach, he flew off westwards.

Taplin, who joined the flight on 4th July, had a memorable experience late in the month that was funny to everybody except Tap himself. The incendiary bomb we were using was roughly egg-shaped with a couple of fins on it, and about 18 inches long. It weighed 50 lbs and was dangerous to handle, and in consequence utilised several safety devices. As usual with this class of article, it sometimes went off when you didn't want it to, and frequently did not explode when it should have. This latter failing was particularly aggravating after you had dodged "Archie" field guns, machine guns and enemy aircraft and dived almost into an enemy hangar to set it off. It was filled with phosphorous, plus some explosive that was supposed to detonate and scatter the phosphorous on impact. On many occasions all the trouble went for nothing. Tap resolved that he would make sure one went off, so he took every safety device off that he could, and made some other adjustments. Then he and another chap, Oscar Ramsay, who had treated another one similarly, started to leave about an hour and a half before daylight with the intention of laying the eggs in the hangar of some unsuspicious Hun squadron. They also carried a 25 lb Cooper bomb, but Fate was to intervece. About half way through Tap's take-off run, he ran into a rut and broke an axle causing, the machine to slew around in a circle and finished up on its nose. Of course the incendiary bomb immediately exploded, setting the 'bus on fire, but Tap, with memories of the deadly Cooper, was leaving for points north "per bootem" as hard as he could go. He had covered about a hundred yards before the Cooper went off, and even then he was lucky for pieces of this type of bomb, "daisy-cutter" frequently went beyond a hundred yards. Although his leather flying coat was pitted with phosphorous burns, he was untouched. In the meantime, Ramsay, who had taken off a few seconds ahead so as to avoid a collision, saw the explosions and the fire from on top, and was chasing around in the dark to find the enemy who had dared to attack our aerodrome. Failing to do so, he returned to see what he could do to help us on the ground, and met Tap just returning from his cross country gallop.

I had another fourteen days' leave due now, so handed over the job to Bo King. Malley saw me as far as Boulogne and promised to meet me on my return. This time I went to Scotland, but did not get further than Edinburgh. In company with two Black Watch officers whom I met on the train, an American major and another Australian named Smith, I saw the sights of the town, such as they were. There was a Japanese commission of some sort staying at the same hotel, the "Royal" in Princes St., and they were an extraordinarily pleasant crowd. There were about twenty of them and they were delighted to be in everything that we did. They had extreme difficulty in understanding the mixed languages that we spoke between us, but enjoyed our company and the only time I ever noticed them in difficulties was when we "pinched" a couple of hansom cabs from a cab rank, and with the

It seems fitting to place a photograph of this big genial fighter pilot, 'Bo' King opposite those of his No. 2 Squadron contemporaries. Here King, with an unknown friend, prepares for a patrol as flight leader. The leader's pennant flutters on the left hand edge of the picture, appearing as a shape on the bottom wing. With so much experience within "A" Flight, it was not uncommon for pilots other than Cobby to lead.

American driving one and an Australian in charge of the other inspected the town. As far as I know, there is no law to prevent officers in uniform taking the air in this manner, but we attracted a lot of attention. Perhaps it was the fact that there were so many of us in the vehicles, and of such pronounced difference in nationality, that we got into the public eye. Anyway, when we saw several mounted police in line abreast moving sedately down the street towards us, we thought it time to dismount, so leaving the cabs in the care of the two or three Japs who were with us, we left hurriedly via the front doors of one of the buildings on Princes Street and then ducked straight out the back. The Japs found it hard to account for their presence in the cabs, and the limitations of broad Scots and Japanese as a medium of discussion did not assist their explanations. The climax to the Edinburgh trip came when a challenge was issued that I could not climb the Scott Memorial from the outside. The monument of remembrance is of yellow sandstone, and rises from Princes Street, opposite the Royal Hotel in the form of a kind of church steeple. To the encouragement of the rest of the team I managed it fairly easily. They were watching the proceedings from the lounge of the hotel on the first floor. Quite a large crowd were interested too, but some of them were not quite as enthusiastically, especially a number of police who seemed to regard it as a sort of sacrilege. Unfortunately, after reaching the top, I could not bring myself to descend and after about an hour's clinging to the top, was ignominiously brought to the ground over the shoulder of a member of the town fire brigade, who had been called out with some of his mates to run up an extension ladder. The crowd was by this time about ten thousand strong, and Royalty could not have received greater attention than I did. When they saw that I was a member of one of the far flung dominions, they treated it as a humorous event, but the police had other ideas, and even the solicitations of their "brither Scots" the Black Watch officers, could not make them see anything funny in it. They were most respectful but very firm, and we were instructed to remain in the hotel until we received a call from the Commissioner. He arrived just before dinner, and without reference to what had gone before, invited us to partake of refreshment with him. He then proceeded to issue instructions covering the next twenty-four hours or so. We were to pay our accounts up to breakfast the following morning, then pack our bags. The two Australians were to catch the next day's train back to London, the American, who wished to go on a tour of Scotland, was informed when his train was to leave and he would be seen off on it, and the Scots were sent off to their homes somewhere north, the same evening. There was no allusion to any of our exploits during the past week, just an insistence that we do as we were told. After one or two whiskies and sodas all round, he left. He had a great understanding of human beings, particularly soldiers. Next day Smith and I caught the London train, and before it left the Commissioner turned up to say goodbye, "being in the vicinity" as he said. During the few moments before it moved off, he mentioned that he had lost his only two sons early in the war and that he well knew they would be doing similar things if they had kept going, but he also mentioned that they would not be doing so in any town whose good behaviour he was responsible for. As we said cheerio to him from the window, he slipped a half bottle flask of whisky through to us, saying, "You might feel like a drink presently." For some time Smithy and I felt like two naughty children who had been reprimanded, but after sampling his gift we felt better. I finished the remainder of my leave in London in my usual sedate manner.

Again I turned my footsteps to France and the squadron with almost the same eagerness one feels when approaching home after a long absence. One seldom became morbid or depressed during those days and only very occasionally would it occur to one that all life held was continual fighting, until at some distant date one went the way of all things, or the war would end. The latter was a very remote possibility, the former ever present as a probability. Everything had gone on well in my absence and I slipped back easily into the daily routine, hardly being aware of having been away.

The Last Months

This SE clipped trees on the boundary at Reclinghem. The pilot is unknown. Just visible outside their Bessoneaux hangars is a line-up of No. 2 Squadron Camels.

CHAPTER 8

One of the doubtful privileges which we were granted in No. 4, was that of looking after the machines both of the Brigade Commander and the Wing Commander. Whether they considered our men did no have enough to do, or whether they thought they would be better cared for by us pagans, I do not know, but as they remained with us I must assume that they were at least satisfied. Brigadier Ludlow-Hewitt, was an exceedingly tall and slim chap of quite youthful appearance and he used his Camel mainly for visits to the different units under his command. Sometimes he went over the lines with the larger formations, but his job was a busy one and he seldom had time to fly. At that time one's impression was that he was somewhat satirical and abrupt in his manner, but he gave us a good run for our money. Apparently he was always in his most pleasant vein on the occasions when he visited our Australian units, because most English squadrons in the Brigade held other views regarding his temperament. Anyhow he ran the business well and other than on one occasion when Malley and I dined somewhat freely at Brigade and tried to put over stories of the Murray cod that we caught in Australian rivers, we always felt at ease with him. He simply refused to believe that there were such things as 100 lb fish in any inland river. Ten years later I showed him photographs taken on the Murray, which I carried to England for the express purpose of convincing him. Unfortunately he had forgotten the incident and was only politely interested.

The Wing Commander, Colonel Louis Strange, was in a different category. He was one of those likeable energetic personalities and no matter how busy the job was of administering the Wing, he could always find time to dash along to the aerodrome, and demand his machine. My greatest complaint, and a perpetual one at that, was that no matter how much fighting we had been doing in the squadron, he was always eager that I should pick one or two other chaps and take him over the lines to have a look at this, that, or the other. He was one of those persons with more than the average issue of guts and a mediocre showing was not to be thought of. All he wanted was to be let loose in any old sort of trouble at all, but we always had more than a stomach full of scrapping in the ordinary run of events. To have an extra show wished on to you at perhaps three in the morning, whilst exhibiting an admirable example of courage and determination, brought many muttered curses on his head. He has since written of some of his flying experiences and in his book "Recollections of an Airman" wonders "what Cobby used to do when he used to get up that hour or so before dawn and disappear over the lines". That, however, was part of my job and amongst other things I used to devote a lot of time to studying where one met with most "hate" so that Strange would get a bellyful next time he wanted to go out! He was a grand type, both as a man and a leader of flying people.

One of the most amusing events I ever experienced occurred on the occasion of my return from leave. A flight commander, whom I shall call George, had come up to Boulogne in the tourer to meet me, and at the same time to takc another chap on leave. In accordance with ancient custom, the other fellow had left a day early in the hope that he would be able to fluke a trip across to Blighty and gain the extra day. George had met me at the boat and told me the Town Major, a cheery soul, had introduced him to a French family who would make his short stay pleasant while he waited for the car to return. Some four hours or so later, about 1 a.m., Boulogne was subjected to one of the Hun's periodical air raids. All street lights went out, alarms sounded, and presently the bangs and explosions of bombs and anti-aircraft guns informed us that the show was on.

The car was outside the club, so hastily gathering up my kit I beat it for the Rue de Renard to pick up George, and found to my horror that his friends' house was on fire, and that the one next door (a terrace of three storied dwellings) had taken a direct hit with a pretty big bomb and had practically disappeared. The Rue de Renard was a very narrow street and was full of gesticulating French people. The local fire brigade arrived and got jammed up against the back of our car, and every time the driver would get away and try to clear the fire engine, the impetuous Frenchman at the wheel, with a kind of "view hallo" would crash into the back of it again. Ultimately they extricated themselves and after vociferous "allez vousing" they went into action, but it was hopeless.

We found George at the end of the street, safe and sound, with his coat and field boots on, but his breeches were over his arm, whilst his nether portions were swathed in a blanket. We parked him in the back seat under the big angora goatskin travelling rug and made out of town. He told us his experience as we went along. "It was a good night," he informed me. "You missed a bonzer party. I was entertained with true French hospitality and drank buckets of bubbly, sang all the songs I knew, played the gramophone and generally enjoyed things."

"They suggested that I might like a snooze and should stay the night and showed me upstairs to a bedroom on the top floor. I heaved off my field boots and my pants and crawled into bed, and presently faded gently away. But it seemed that I had hardly fallen asleep when something went off with a terrific bang. I could hear whistles being blown and people crying out "Le Bosche!", "Beaucoup Bombard!" and so on, but I was quite comfortable and much too sleepy to move. Then something like ten thousand tons of bricks feel on the house next door and the next minute I was looking out into space. The wall had disappeared and so had the house next door. Recollecting something about lightning never striking twice in the same

This SE5A was engulfed by fog when returning from a dawn patrol. It ran out of petrol and landed heavily with a dead engine.

Lieutenant Frank Currie beside his "C" Flight SE5A. A protective canvas covering had been fitted to the Lewis gun on the top wing.

place, I settled down again. I awoke with a choked feeling and sitting up I saw the room was full of smoke and sparks. The remains of the house next door were on fire, and it seemed that the lower floors of the house I was in were also alight. Grabbing my belongings and wrapping a blanket around my nakedness, I rushed downstairs, but had to put my field boots on to cross the lower flight as it was smouldering steadily. I made for the end of the street hoping to get a lift around to the club, and apparently missed you." With this explanation George went to sleep under the rug, wrapped up in his blanket. He was still there when the car came to a stop on the tarmac at Reclingham outside the orderly room. The Adjutant, or Recording Officer, was the only officer on parade and he had just wheeled the men off the "falling in" ground and they would presently pass us. We rarely had a parade in the squadron. The nature of our work demanded that small flying parties would work shifts according to what machines were flying, and as these were continually coming and going, no more than a series of squads of eight or ten men would fall in outside their tents and march straight to their duties.

Apparently the weather having been too dull for early flying, the men had all come on parade together and the pilots, true to type, were sneaking an extra bit of shut-eye.

Old George, who knew nothing whatever about drill, and could barely give a command correctly, was nevertheless a stickler for the right thing, and he too had obviously just awakened. As the troops came opposite the car, the adjutant gave them the "eyes right" and slung the appropriate salute in the direction of the car. George, befuddled by sleep, but nevertheless true to his principles, got quickly out of the car and drawing himself up rigidly to attention, returned the salute. He was greeted with an expression of blank amazement on the fact of the adjutant, then someone tittered and in about a split second the whole crowd were laughing hilariously. George was astounded until the corporal driver leaned out of the front-seat in a hearty whisper informed him, "No pants on, sir."

George ducked back into the car with his khaki shirt-tail blowing in the breeze, much more quickly than he got out.

July proved to be our most successful month while in France. We shot down 38 enemy aircraft and our aircraft which only totalled 25, crossed the line 610 times. We dropped nearly 7½ tons of 25 lb. bombs, fired 40,000 rounds at ground targets and another 7,480 in air combat. On one day the total of hours flown by the squadron was 86 hours 5 minutes. This was a month of very solid work. In addition to L.E. Taplin, D.C. Carter and M.H. Eddie joined the flight and were taken in hand, whilst Armstrong and Scobie returned to the R.A.F. They had been on loan during a shortage of qualified Australians.

August 6th was the day of my first show over the line after this spot of leave and it started with a visit by Trescowthick and I to the area just behind Veille Chappelle. We obtained direct hits on a dump and started a series of explosions. A fire started to blaze away merrily. About three minutes later we saw a train on the light railway running frome Veille Chapelle back into Lille and put in some thousand rounds or so into it from two hundred feet and dropped one bomb square in its middle. There were no enemy aircraft around and after chasing along the trenches for a while, we returned home and made an arrangement to go out again later. As it turned out Tres could not leave the ground owing to a faulty engine, so McClaughrey, the C.O., Bo King, Watson and I went out about eight in the evening and had a "look see". We immediately spotted a dark, weirdly camouflaged L.V.G. over Bac St. Maur, and I pounced on it from behind, opening fire when close up to it, and continuing on until I was under its tail. It did two turns of a spin after starting off with a wobbly sort of sideslip. It again commenced to spin, and I followed, firing from close range until it

Lieutenant E.A. Cato of No. 4 Squadron in the cockpit of his Camel. On the cabane strut is a wind-driven generator to pressurise the fuel tank.

crashed south-east of the bridge at Bac St. Maur. King had remained above me to guard my tail from any enterprising sportsman that might have had designs on me, then the C.O. and Tres dropped four bombs on some troops and transport that they noticed on the Estaires Sailly Road. As I rejoined Bo, they were stunting just off the ground a mile or two away.

Tres had his machine OK again by the next morning, so we left in company at twenty past eleven on a special mission. We slipped through the clouds towards Armentieres, and when somewhere over the lines, put our noses down and dived through them with the intention of getting rid of our bombs as soon as possible. We came out into a clear sky at three thousand feet, and found ourselves behind five Pfalz scouts flying north a few hundred feet below us. I pointed them out to Tres who was alongside me, and whilst still in the dive, released my bombs and turned to come at them directly from behind. My opening burst was immediately successful, for the machine I was attacking burst into flames immediately. Tres's opponent had dived away as he saw my companion on his tail but had given him a comparatively easy shot from behind. This machine also caught fire. A third Pfalz had also tried to get away by diving and was some distance from us and well below. I dived on it from the front and above as it banked round towards its own territory. I fired about two hundred rounds altogether from three hundred feet down to about twenty feet away, and as I passed over it, it was smoking from the centre section. After a few moments it slowly went into a spin and burst into flames when it had almost reached the ground.

We then had a change of scenery. After the big offensive commenced by the Allies near Amiens on 8th August, the squadron was ordered to fly south daily to this area to reinforce the air units there. Fortunately it had only lasted for three or four days from about 10th August, but things were pretty willing while it lasted. The job necessitated long flying hours each day, as on top of the offensive patrol duties, we had the long flight down and back. Just before the crack of dawn, and it seemed to crack very early, we would leave Reclinghem for Allonville, a distance of some hundred miles. This meant a scratch breakfast at 3 a.m. and on arrival we would fill up with petrol and oil, put the bombs on the fit ammunition belts and go straight off on squadron patrol. Horseshoe Wood, our temporary aerodrome, had no facilities for refuelling or serviceing the aircraft. The first day we existed from our 3 o'clock "duck's breakfast" until we got back home again at about ten at night. Thereafter we took enough down with us to keep us going, but as we had only single-seat aircraft, we could not carry any mechanics to look after our buses. Worse still from our point of view, we had no one to prepare any "grub" for us, or to even look after our sandwiches while we were away. There were a few stray mechanics hanging about, but as there were about 70 or 80 aircraft temporarily using the ground, one's chance of assistance was negligible. It seemed to us a poor way to win the war.

This short period was one of the most inspiring experiences of the war. Apart from the successful gigantic air fights we had in formations of anything up to sixty or seventy machines against somewhat similar numbers, the sight to us for the first time, of Allied troops "putting in the boot" to no mean order on the ground, gave one a feeling that the war, after all, might finish within a reasonable time. Roads were packed with column upon column of prisoners on their way back, and our advancing troops seemed literally to leap across country almost without interference. I shall never forget the sight just in front of St. Quentin about the time Peronne was taken, of our field artillery galloping into action on a low rise, well in front of the infantry, with the enemy gunners in clear view in front of them, limbering up and galloping into the protection of a nearby hill. Success undoubtedly inculcates success, even if one is only indirectly concerned with it and we definitely felt on top of things in the air. Nothing seemed too big for us to tackle.

We were again visited by Australian war correspondents at Reclinghem and resolved in some way or other to even up the score in a mild way, against journalists in general, for the drinks we had had to buy owing to the many over-colorful descriptions of some of the jobs we had done, and had appeared from time to time in the home press. Those of us who were still in the squadron, remembered the teasing that Tab Pflaum had had to put up with as a result of the many columns that had been devoted to the "Baptist Boy Hero" and certain of his exploits. Then Garney Malley had been covered as the "Mosman Air Marvel" and in long articles, episodes of his early life were referred to, how he used to intrepidly breast the surf at Manly at an early age, as he was now breasting the heavens!

As for me, they had got hold of odd letters that I had written home to friends and so on, and quoted extracts from them as though emanating from me for direct publication. Our letters home were naturally written in the most optimistic strain to help keep up the spirits of the home folk — not that they required any aid to do this, as they were all wonderful, but it was part of the game. Then to get columns of stuff back from Australia headed; "With the Australian Flying Corps in France by our Special Correspondent", decribing graphically the daily tasks of the squadron, in themselves quite good subjects for writing, but colored with a lot of hot air purporting to emanate from ourselves, made us sitting shots for everyone who came into the mess. When it appeared under inch headlines, those people who read the stuff must have thought that if bombast would win the war, we should go a long way indeed. All of it came back to the squadron in papers from home. Fortunately we all had such a sound knowledge of each other's natures after living, fighting, drinking and practically sleeping together for months, so we knew to the utmost how we could rely on each other both in the air and on the ground. Such guff did not worry us between ourselves, but there was naturally a good deal of leg pulling over it. What other people thought who did not know us, I can only imagine, but it wouldn't be very flattering. I was always billed as the "Elsternwick Boy Hero" after a suburb in which I had been known.

One of the correspondents was Keith Murdoch, nowadays a knight and managing director of a group of important newspapers. He was a pleasant and cheerful personality and most interested in us all as individuals and always evinced a large amount of wonder and admiration in our harebrained escapades. Personally he probably had nothing to do with the "guff" but we decided that the unsuspecting Keith should be the lamb to be led to the sacrificial altar for war correspondents generally. He was rather keen to be taken up to see the lines from the air, but as our machines were single-seaters we couldn't do anything in our squadron, but No. 2 had a stray Sopwith Two-Seater that they had acquired by rather dubious means. It past was veiled in mystery and how a spare Two-Seater came to be held in a fighter squadron of single-seaters was a secret to most people. It was understood that the French Air Force had suffered a deficiency of one

Lieutenant G.H. Blaxland and his air mechanic (name unknown) pose before their "B" Flight SE5A. This aircraft carries a particularly interesting set of white markings, the style of the figure 2 being quite unusual.

No. 2 had a stray Sopwith Two-Seater, its past veiled in mystery. "It went by the friendly name of Sophie."

aeroplane some time before. It did not possess any guns or other war equipment but was used for joy riding and paying friendly calls, and even made unobtrusive visits to Blighty for brief periods. It went by the friendly name of "Sophie".

Malley and I arranged with Follet, one of No. 2's flight commanders, to take Keith up, but we also arranged on the quiet that he should go well over the lines and throttle off his engines for as long as possible, so that the Bosche "Archie" could have a good go at him. For the information of those who don't know, a burst within a hundred yards is a loud cough, and the closer it gets, of course louder the noise. When it is a decided "bang" it is close enough to lift you out of your seat if it does not blow you to pieces, but the sound to the man in the air is never as loud as, say, a similar burst to the man in the trenches. The noise of wires and engines smothers some of the sound and further, the flying helmet and perhaps cotton wool in the ears, assists in muffling the noises. Again, unlike the man on the ground, you cannot hear the shell as it travels through the air, unless you shut off your engine, and then due to the silence of everything up top, you hear everything — plus. The cough becomes a frightful "crump" and the near misses just about go right through your head.

Malley and I were to escort Sophie in case any sportive enemy pilot should become interested, but we were to stay well up above out of the way until Murdoch had had a good introduction to Archie, and then we proposed to give him three pennyworth of "dogfight" during which we were to demonstrate an aerial combat to the passenger.

Everything worked according to plan, except that Follet nearly crashed taking off. The leg reach to the rudder bar was particularly long in machines of the "Sophie" type and as Follet was built on small lines he could not get enough leverage on the rudder to stop the bus from swinging as he started off and it described a big circle. He knocked over some crates of 25 lb. bombs on the tarmac and nearly collided with the end hangar, but it got off all right in what the press used to call a "clambering swerve".

On the way to the lines, we poked our wingtips in alongside the ear of Murdoch in the rear seat, to let him see how safe flying was, and perhaps to inspire in him a little nervousness in preparation for "Archie" before crossing the lines at about twelve thousand feet, quite a nice height for effective shooting. Of course Follet was expected to use all his gumption to avoid being hit, but there was of course a fair amount of risk.

As soon as "Archie" opened up, Malley and I climbed into the gallery to enjoy the spectacle and presently we saw "Sophie's" nose come up slowly as Follet shut off his engine and gently stalled the old bus. Then he pulled it up to proper stall and held it there until a tail slide was imminent, before letting it fall over sideways and recovering control about four hundred feet lower down. In the meantime, Archie was giving them a good going over and the air was black and brown with bursts of high explosive. After about fifteen minutes of doing everything one shouldn't do for bodily comfort, Follet made back to the lines and then Malley and I in our Camels staged a mock battle with the Two-Seater.

We dived straight at it until we could see Murdoch duck his head as we shot across him, and took it in turns looping around the machine, rolling alongside it and then closed on it again. We then flew sedately homewards making signs to Murdoch appropriate to the occasion. Back at Reclinghem we asked him how he enjoyed things and what kind of chance he thought we had of finishing the war in a blaze of glory as his paper had so often inferred. He took it all extraordinarily well and though he looked a bit pale around the gills, insisted that we adjourn to the mess and have one on him. Naturally we duly honored the invitation as we rather liked his style and later came to value his friendship.

George Malley then left the squadron for Home

Around the corner of a Bessoneaux hangar, and from a ladder one "Sergeant Clemens of Melbourne" photographed this two-squadron line-up on the occasion of a visit of Australian General W.R. Birdwood. The Camels are in the background and Cobby's aircraft, A, is partly obscured by the hangar guy rope.

Establishment. We had been together from the start and had seen the fighting strength of our unit turned over many times. We had been in practically everything of importance that had occurred, both on the ground and in the air, and his leaving was a wrench. So far I had been lucky enough not to be sick or wounded, but he had had a couple of knocks. In March he had received a bullet through his leg, but had kept on going and on another occasion, had been hit by shrapnel. I was wondering how long it could last. But the fatalistic outlook that had become so much part of our natures, just refused to be concerned with the possibility of being knocked out of the sky. One's great concern was whether one was to be confined to endlessly flying and fighting, and whether there would be any other sort of existence for us. I saw Malley off from Boulogne in the accepted and time-honored custom.

Three days later on 10th August, King took the principal part in things. While I sat over him, he attacked a balloon just east of Estaires, and fired at it all the way down in a long dive. It started to smoke soon after he opened fire and in a couple of seconds was well alight. As he came up to rejoin me, I was attacked by seven Albatross scouts who dived from above, and I immediately started to throw my machine around as much as I could. I had a couple of snap shots at one or two of them as they whizzed past me, but without any effect. For some minutes I was fairly busy, climbing, half-rolling off the top of loops and doing everything conceivable thing I could do to ensure that none of them would have time to get his guns on me. I was not concerned with attempting to get a shot at any of the Huns, unless one got in my way and even then only fired a very short burst before pulling away. I kept an eye out to see what was happening to Bo, but he did not turn up. I discovered afterwards that he had his hands full with several others that had arrived quickly from behind some clouds. Fortunately a formation of S.E.5s now took a hand and in a few minutes my late opponents were heading for home with their heads down and the S.E.5s after them. We had only been flying another few minutes, when we saw an S.E. having a go at an L.V.G. Bo went down to lend a hand, the S.E. pulled out in a zoom and cleared off to our side of the line. He must have thought we were the enemy. Bo opened up from in front of the two-seater, which immediately started going around in a circle. It is difficult to hit a machine when it starts doing this, and Bo had all his work cut out to get into position. Finally he again got in front of it, and fired a long burst from close up, then just managed to get out of the way without colliding with it. The L.V.G. went into a steep dive and came out lower down, with the pilot apparently wounded. He tried to land but overturned and crashed.

That some afternoon we were granted the unusual honor of a visit by His Majesty the King. We had already had had the privilege of a visit by H.R.H. the Prince of Wales, but to the members of the two Australian squadrons so far away from our own country, the presence of the King on our aerodrome was an event that was sincerely appreciated. Most of us had never seen him before and the informal manner of his conversations, and his expressed desire that no interference in the work of the squadrons should be caused by his presence, made us feel that apart from any other qualities he possessed, he was a very pleasant English gentleman.

He was particularly interested in the scrap that Bo and I had had that morning. We were surprised to discover that he possessed an excellent understanding of air fighting, and that our general allusions to the fight were not sufficient and he required quite a detailed description. "Birdie", otherwise General Sir William Birdwood, the G.O.C. 5th Army, over whose troops we were working, was in attendance and, of course, the Brigade and Wing top brass.

I then made another personal visit to Wavrin aerodrome. For some time I had had ideas of being able to set fire to the enemy hangars when all the machines were inside. To do this it would be necessary to get over well before dawn, or late at night, or on a day when the weather would be too bad for flying. My idea was to take over one fifty pound incendiary bomb and a couple of ordinary Coopers, start a fire, and then with my machine guns force everyone to keep out of sight so as to give the fire

A very significant picture. His Majesty King George V, visits No. 4 Squadron at Reclinghem on 10th August 1918. General Birdwood looks at the camera as Roy King and Harry Cobby converse with the royal visitor. Behind His Majesty is Major McClaughry. The line-up of pilots is important, particularly as the first on the left is the only available picture of Lieutenant Norm Trescowthick. Next are Lieutenants R.G. Smallwood, L. Bayer (Armament Officer), R.R.B. Bennett (Recording Officer), D.C. Carter, R.T. Edols, M.H. Eddie, and, half obscured, L.E. Taplin. Within days Smallwood and Edols were wounded and within a month, and on the same day, Carter and Edols were killed in action. Taplin (12 enemy aircraft) was shot down and captured, also on the same day.

every chance of spreading. I obtained a direct hit with two of the bombs, but the incendiary one did not go off, so I tried to start a fire with my machine guns, but again I had no luck. I gave it up and came home through the rain and clouds feeling that I had wasted my time. This was about the fourth time that I had made an attempt to wreck an enemy aerodrome, but without success. The squadron, however, was to experience the pleasure very soon.

For some time we had been warned that a raid against an enemy aircraft park would be carried out as soon as the opportunity offered. It was to be a Wing affair, and all available aircraft were to be employed. This meant up to eighteen machines from each of No. 88 Squadron, R.A.F. (Bristol Fighters), No. 92 (S.E. 5s), No. 103 (D.H. 9s), No. 2 Australian (S.E. 5s) and No. 4 Australian (Camels) would probably be employed. Preliminary details had been arranged in several conferences. I was to lead the entire wing to its objective and launch the first attack with our squadron, followed by No. 2 Australian, then the other Camel squadron, with the Bristols keeping guard above us, while we went right down on top of the roofs of the enemy hangars and attacked from really close quarters. The people on top would then attack when we had got out of the way. The target was not to be decided until such time as the authorities had collected sufficient information to indicate that the drome selected possessed a large number of machines. Personally I felt a little apprehensive about taking such a large number of aircraft across in one formation, as even the most indifferent shooting by the enemy anti-aircraft gunners should be able to knock some of us out of the sky. The arrangements were that we should all move off from our 'drome, Reclinghem, at a given time, this time to be withheld until the last possible moment. Each squadron had a fixed time to arrive over the 'drome at two thousand feet, with all their aircraft in "V" formation. That is, each flight would be in a "V" with either five, six or seven aircraft in each of the three flights, and each flight would again form a "V" with its leading flight in front and the other two behind and above and on the flanks of the leader. They would then be echeloned in height and to the flanks of the leaders. As they arrived over the 'drome, they were to move around the outside perimeter of the ground in formation. This meant that they would be moving round a large rectangle. As the second squadron arrived, the first one would move up a thousand feet and both would continue to circle the drome. Then No. 2 Australian Squadron, on the ground waiting to take off, would move off in flights, form up in the air and carry on with the movement while the others went up another thousand feet. When they were all off, we in No. 4 were to move off in similar fashion by flights, and take our place, in our case, the lower position. Then I was to lead for the line with my flight in front, flanked by "B" and "C" Flights, each one slightly behind and above my level. No. 2 Squadron followed in similar formation, with No. 92 behind them again, then No. 103 with No. 88 bringing up the rear and holding the highest position. From the ground we would then look something like the backbone of a fish. I was to cross the line at 7,000 feet with the highest and rearmost machines at about 10,000 feet. Then I was to make straight for our objective, wherever it might be, aiming to finish down near the rooftops. As all the enemy 'dromes were located twenty miles or more over the line, we would have a good speed up by the time we arrived at the appointed spot.

Our arrangements had stood like this for several weeks, and other than one or two discussions to make sure that everyone understood his appointed tasks, nothing further eventuated until I was advised late on 15th August, that the job was on for the next day. I had asked particularly that all aircraft should follow each other in single line when dropping their bombs so as to avoid the risk of hitting our own people. The importance of this air discipline can be realized when it is mentioned that a number of machines attacking a single line of hangars, would be running into one another if any independent action was followed, quite apart from the danger of dropping an odd bomb or two on their pals who might be below. We were then warned to be ready to hop off at dawn the next morning, and that orders would be issued during the evening. I was still worried about the risk of losing several people through "Archie." Up until now I had not lost a man over the lines and it seemed to me that unless we could do something to stop the enemy aircraft, we stood a good chance of losing several. I had frequently noticed over the line, that while fighting aircraft always got "Archied", Army reconnaissance machines could fly up and down at varying heights with impunity, and the thought had repeatedly suggested itself that there was an element of live and let live in the business. One of the jobs of army co-operation machines was to turn our artillery onto any enemy batteries that they could find, or onto any other targets of importance that presented themselves. Consequently they were continuously watching enemy territory. Obviously they would know the positions of the enemy anti-aircraft batteries, but unless they proved troublesome to our Army co-operation chaps, they would probably be left alone. In any case they normally did not present a sufficiently important target to waste the time of big guns. The next day, however, was going to be an important one in my estimation. Consequently I rang up some friends in these squadrons, and asked if they couldn't switch some of our artillery fire onto their positions in the area concerned, as we approached or were crossing the line.

In the morning at the appointed time it was raining, so the show was put off until the weather cleared. At half past eleven, however, word came through that zero hour was twelve and our objective was Haubourdin aerodrome, part of the command of quite an interesting personality, von Leutzer. Von Leutzer had a somewhat similar command to that of our own wing commander. He was in

Haubourdin aerodrome under attack during the big raid referred to in the narrative. On the extreme left of the original photograph, Cobby's Camel could just be seen returning to attack. The upper surfaces of its wing and tailplane had been painted white for ease of identification. Burning hangars and aircraft were responsible for the smoke.

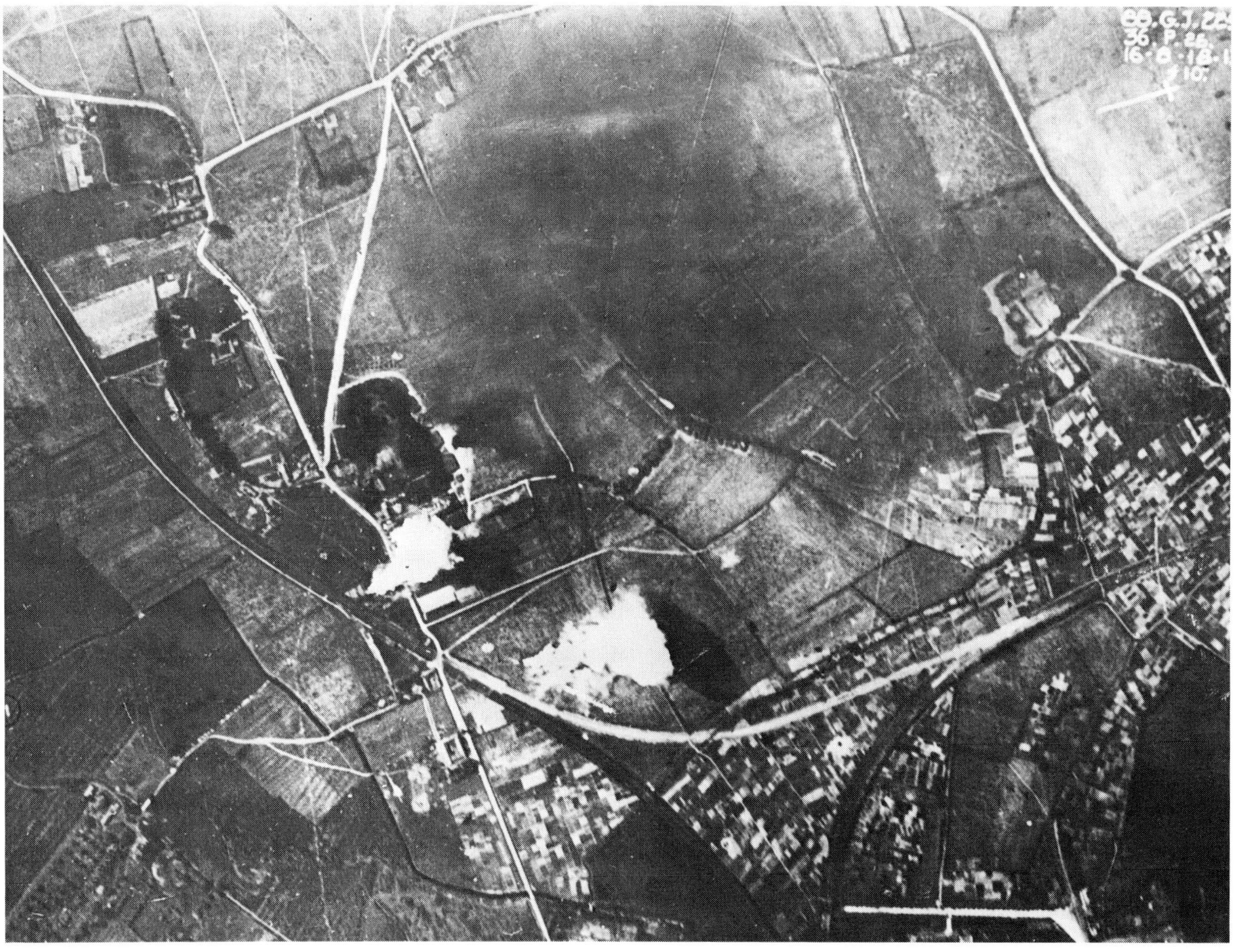

charge of the fighting machines opposite us and his pilots were the chaps with whom we had been scrapping for some months. From all accounts he was an efficient chap and those who had known him before the war liked him.

We were now about to move off and have a go at him on his own ground. There would be from seventy to eighty machines in the attack, and if my private arrangements against the enemy "Archie" worked, the show would provide a pleasant interlude.

We were warned that zero hour would now be 11.45 and promptly at that time No. 92 Squadron S.E.5s, appeared over Reclinghem in "V" formation at about twelve thousand feet. Soon after the Bristol Fighters of No. 88 arrived and as they moved towards the 'drome, No. 92 Squadron went up. Then came No. 103 with their D.H.9s and they moved in under the Bristol Fighters. In the meantime, Murray-Jones, commanding No. 2 A.F.C., was taking off from alongside us and the whole party in the air was moving in the same direction at their varying heights around the outside of the 'drome. Then my first flight went off led by McClaughry, closely followed by "C" Flight under Gillis-Watson. Then I moved off with "A" Flight, and as the formations in the air saw my white top wing leave the ground, they swung around to take up their positions above and in the rear. (I had had the upper surfaces of my wings painted white in contrast to the ordinary drab khaki colour, so that those following could easily pick my machine out.) We made straight for the line climbing slightly all the way, so that I should cross at the pre-arranged height of 7,000 feet. Colonel Strange came and flew alongside me and stayed there until I made my dive on the hangars. We crossed over just south of the Forêt de Nieppe near Calonne, and I hoped the troops in the line, on both sides, were duly impressed. It must have been an inspiring sight from the ground to see the long herringbone formation on its way, but the nature of the inspiration would depend upon which side you were on.

Leaving the throttle open, I immediately put my nose down and made directly for the southern corner of Lille which was close to our objective, Haubourdin. It was nearly thirty miles away, but the aerodrome would not be recognized from this distance. I judged the angle of approach to bring me out a couple of hundred feet up as I crossed the aerodrome boundary. I intended to then steepen my dive so as to run away from the others to allow them to drop into line behind me. Looking back over my shoulder I got a tremendous thrill from the sight of everyone keeping perfect formation. We were impregnable as far as an attack from the air was concerned and the only danger would be "Archie" from Lille and machine-gun fire from the aerodrome guards.

About half way over we had a lesson in what fear can do to a pilot. A German two-seater appeared ahead and below going in the same direction, then down went his nose with all of us after him. If he had turned away he would have been perfectly safe as we would not have wasted time on him, but he must have had one foot on the stick and the other on the throttle, with his head probably nearly screwed off watching us. He just kept merrily on about two thousand feet in front of us, until he suddenly flew straight into the ground just short of the aerodrome.

I had my Verey pistol in my hand, and firing a red light to let everyone know the game was on, I put my nose further down and made for the nearest hangar. The wind was going to be helpful as it was blowing down the line of hangars. I let my first two bombs go from about fifty feet then circled out round the drome to come back along the tarmac to shoot up several Fokkers standing unattended. Bo King suddenly appeared on my left, obviously with the same intentions. He tried to jockey me out of position, but I wouldn't give way so he dropped in closer still and a little behind, and we sprayed the Fokkers with incendiary ammunition until they caught fire. It was too dangerous near the hangars though, as there were still about sixty machines above us, all with four to eight bombs apiece. They were now whistling down all around us, so we turned out across the ground and came back firing through the open doors at the machines inside. Several caught fire, and it looked as though the whole show was going up in smoke. What with the two hundred and fifty odd bombs we dropped, combined with the smoke from burning hangars and machines, and mingled with this the grey smoke of our incendiary ammunition, the result was quite impressive.

As we zoomed over the hangars, several bombs from our enthusiastic friends higher up burst just alongside us, but other than getting a bit of a jolt, no harm was done. Nevertheless from then on we kept outside the danger zone until everyone had laid his eggs.

Bo found a squadron of cavalry coming out from Lille at a hard gallop, but it didn't stay an orderly squadron for long. Some of the other chaps who found it too risky to go in a second time owing to the bursting of bombs looked for other targets. A couple of them found what was obviously the officers' quarters and they gave it plenty of attention also. Trescowthick, I think it was, devoted bursts of machine-gun fire at a big touring car in front of the building, When we got back he said he thought the owner would find it a little hard to start.

Oddly enough we had a glimpse of what courage can do, as opposed to the result of the fear-stricken effort previously referred to. While the whole crowd of us were mixed up over the 'drome, an old L.V.G. two-seater barged right into the middle of us, the observer standing up and letting everyone in sight have it as fast as he could, but hitting no one. I went up under his tail intending to finish him off, but found I had no ammunition left. I tried to attract some of the other chap's attention, but could not do so. After a few minutes the L.V.G. calmly pushed off and was not interfered with. I found out afterwards that several other fellows saw him, but they also had run out of ammunition and could do nothing either. Apparently those above mistook him in the turmoil for one of our own machines, but anyhow good luck to him! One had to be either very brave — or mad — to voluntarily attack a crowd of our size.

We reformed and moved off without hindrance, and I think that our orderly and unopposed return must have had a more depressing effect on the enemy troops down below, than did even our attack on their aerodrome. Photographs taken during and after the raid, showed the damage done had been considerable. The aggressive spirit certainly produced a satisfied feeling in us. It had been a most successful job and we heard later that some sixty aircraft were destroyed.

That night we had an impromptu inter-squadron binge throughout the wing. We all mixed and each mess threw a party. At 4:30 a.m. I was in No. 2 Squadron with Murray-Jones sorting out chaps to the various tenders that were to take them home, when Strange rang up to say there was a similar show on the same morning and not to make a night of it. We were to leave at seven — two and a half hours away, and as far as I could judge most of the

The bombing raid on Lomme. The picture shows rows of hangars on fire.

80th Wing were scattered around Flanders and exhilarated to some extent or other. We hastily got in touch with all the other squadrons and by about six o'clock I should say everyone had got home to his own unit. You could almost feel the air vibrating from the number of throbbing heads in that particular portion of France that morning.

I undressed, had a bath, lay down in my pyjamas, prayed for a dud morning, and faded away. It seemed I was immediately awakened by my batman who informed me that the Wing had sounded the rally and that another job was on. Being a old dog for a hard road, he had provided a "bracer" and a couple of aspirin and, emulating the duck, I breakfasted on these, pulled my flying gear over my pyjamas, and roused the other chaps. We were a quiet looking lot as we stood on the tarmac watching our buses being wheeled out. I looked up Murray-Jones and King Cole at No. 2 and asked them if they knew where the show was, but they didn't. Nobody seemed to know, so I got on to the 80th Wing and they said Strange had left and should be along presently, but they would send the particulars by despatch rider as soon as they were ready.

Soon the other squadrons came along as on the previous day and took up station, but one could see with half a pilot's eye, and that's about all any of us had. Murray-Jones took off with his crowd, then "B" and "C" Flights of my squadron, and still I, who was going to lead them, had no idea where we were to go. I had visions of taking the whole crowd for a trip along the coast just for a joke, and I am sure they would have pensioned me for life If I had. I was still sitting in my machine twiddling my thumbs, when the colonel and the despatch rider arrived together. I had enough sense to orient the photograph of the enemy aerodrome (which was Lomme on the other side of Lille to Haubourdin) with the map, pick out the direction of the wind in relation to the run of the hangars and then take off. I don't think anyone else in the air, with perhaps the exception of Strange, knew our destination. There was, however, one very fortunate feature in the direction of the wind. It was blowing practically right down the line of enemy hangars, and if my memory served me correctly, the hangars were of wood and canvas and should burn well.

Our track should have been similar to that of yesterday, but so as not to give any warning of our objective, I flew further north before crossing the lines, and then carried on in an exactly similar manner to that of the previous day. Things worked out just as smoothly and perhaps because of the wind, more effectively. The colonel had again tucked himself in just behind my left wingtip, too close this morning for comfort, and I said enough to burn the dope off his wing tip. I made up my mind, however, that if he wanted excitement he would get it. Instead, therefore, of finishing as low as we had the previous day, I stayed up a thousand feet or so and then pushed straight into an almost vertical dive that nearly bucked me through the centre section. He lost just enough distance to allow me to flatten out under him at the last fraction of a second, instead of diving through the roof of the hangar. My first two bombs started fires in the second and fifth hangar. There were six in this row, with another five on the other side. Strange tried to hold me as I turned for the other row, but he couldn't. He had to wait for my turn before he knew which way I was going and that little fraction of time gave me perhaps forty or fifty feet on him. Down along the second row, letting my bombs go without sighting, as I couldn't miss from only about twenty feet above the rooftops, and one hangar started to burn immediately after the bombs had burst. Then back to the first row again, across the centre of the aerodrome and firing as I went, with my wheels almost touching the ground, and Strange doing his utmost to keep alongside. Skidding over the hangars which were burning furiously, with bombs going off all around us, I raced around the outside of the aerodrome, firing bursts at odd groups of officers and mechanics that were making judiciously for the "tall timbers." In about ten minutes there was so much smoke that the danger of colliding with some of the other chaps was too great to stick around, so I climbed up to have a look. The scene was a thrilling one to the eye and although I did not know how many machines were in the hangars, I guessed very few could now be saved. At a rough guess, I should say there would have been about thirty to forty machines destroyed, apart from the odd damage done about the place. To be able to go and smoke your enemy out of his hide-out, however, and then to return with impunity, had a far greater effect upon the morale of our fellows, than did any realization of the actual material damage done. Naturally, the other chaps' tails were down to the same extent that ours were up.

We received our first batch of the new Sopwith Snipes soon after this, but I did not have enough practice on them to be able to handle them properly. "Bo" King liked them from the start, as his rather large frame was given plenty of room in which to move. They were not used over the line before I left for Home Establishment, but soon after I left they gave the enemy some nasty shocks. He had been used to having things his own way higher up, and the first time the squadron met them in numbers, the enemy thought he had some easy pickings, but he was wrong. The first month the Snipes appeared over the line produced a record for the number of enemy aircraft destroyed.

Although August had been a particularly bad month for weather, with rain and storms almost daily, our squadron aircraft nevertheless crossed the lines 684 times.

In addition to Malley's departure, Smallwood was admitted to hospital with a piece of high explosive in his back, and our "Canuck" — McRae — was claimed by the R.A.F. He had played a very good hand and his cheerful disposition had been a great asset at all times. Also of interest was the promotion of Gillis-Watson to command "B" Flight in place of Malley. As with myself in early days, the C.O. demanded many things of the prospective flight commanders that were entirely redundant to the job of leading fighting pilots into action, and Watson had been passed over once before. Sheet merit and ability could not be displaced by any form of smugness, however, so Gillis became a captain and a flight commander.

Massed Patrols

CHAPTER 9

These massed patrols became a regular habit, and several times a week we would combine with other squadrons of the wing and go out and smash the enemy on the ground. On many occasions we met large German formations, and a number of battles took place in which anything up to a hundred aircraft took place. Strangely enough, these huge fights seldom caused large numbers of casualties. The air was too crowded and everyone so mixed up that there was little opportunity to settle down and have a steady shot at anything. You would no sooner pick out someone to have a crack at, than there would be the old familiar "pop-pop-pop-pop" behind you, or you would just glimpse a doughty enemy pilot getting into position to fire on you — so hard boot and stick one way to save your skin. Machines would whiz past each other continuously in every conceivable direction, and frequently collisions would occur. They did not last long, a few stragglers or those that dropped behind the scrap, would be picked off and presently both sides would be moving off in opposite directions.

Early in September I received orders to proceed to Home Establishment, as Wing Fighting Instructor to the Australian Training Wing in England. I tried every conceivable way possible to get out of it but without avail. I even tried to wangle a move to the Italian front or to Palestine, but it was no good. Colonel Watt, the Commanding Officer of our wing, was concerned with the training of pilots on Camels particularly, and was also anxious to push on with more fighting and air gunnery practice throughout his command — so I was for it. The prospect of being chased around the sky by enthusiastic, but only partly trained pilots, was not pleasing. France was dangerous enough, but England seemed more so. The last two fighting instructors had been killed in the air in collisions, and the prospect of passing out in that manner was not inspiring. I managed to obtain a couple of weeks' grace, with the excuse that I wanted to try out the new Sopwith Snipes. They proved to be much faster than the Camel, had a far higher ceiling, and did not possess any of its vices, although we had mastered these long ago. In the meantime, I more or less handed the flight over to Bo King and freelanced a little, although I joined in most of the bigger shows.

With a couple of days to go before I had to leave, we were provided with a very bright entertainment by the 5th Army Concert Party, under the direction of Leslie Henson of musical hall fame. He had a team of first-class artists including a couple of female impersonators, one of whom was Bert Errol. We cleared a hangar and rigged a stage, the "company" providing all the props and drop curtains and so on. They had a pretty good orchestra with them too. Naturally we had a guest night dinner beforehand. General Birdwood, who was commanding the 5th Army, and most of his staff were there. Brigade and wing commanders and staff and odd friends were in strength, and both No. 2 and our own officers' mess was packed with people out for a bright night. They certainly got it. I know most of our chaps swapped jackets and hats with the visitors before we finished, and I saw the concert out in a staff colonel's regimentals. Later on the make-up was added to by Leslie Henson fixing an almost real dragoon-like black moustache to my face.

The troops really fell for the impersonators, particularly a delightful blonde, and comment was most audible and particularly personal during "her" performance. Bert Errol, almost Spanish as a brunette, sang the Indian Love Lyrics in fine voice — and got away with it. But the irrepressible and incorrigible Australian "digger" could not refrain from tendering advice to the partner of the "blonde" in a passionate love scene.

An elderly bishop, Padre-in-Chief to the 5th Army, was prominent in that he retained throughout the evening, three long "sausage" balloons, one red, one white and one blue, that had been arranged in the form of the Prince of Wales feathers. They were articles for the use of the troops and had been the only adornment on the dining table and had been placed over the president's chair. At one hectic period of the dinner, they had been auctioned and although the "Bish" had not made a bid, they were knocked down to him. His refusal to part with them earned him many light, but cryptic remarks upon his mixing the flesh with his ecclesiastic office.

Then tragedy of the worst kind since our arrival in France struck "A" Flight. We had been particularly fortunate right from the start in that we had not lost anyone, except during the first few days. From the time I started to lead the flight we had not lost a man over the line, and now in one patrol we lost four. Bo and I had been around in the afternoon for my farewells to all the people I knew in the vicinity, and he had arranged for Trescowthick to take the evening show of five machines. When Bo and I returned in the early evening, when they were due to return, but only Norman Trescowthick came back. The others had all been shot down.

A tension always gripped everyone on the tarmac on the return of patrols from the line. Its extent varied according to what machines could be sighted coming in. It was the custom to come in over the aerodrome in formation and under no circumstances would the order of flight be broken until the leader gave the signal — usually by using a white Verey light. Then the formation would split up, and the behaviour of the pilots as soon as they fell out was a fair indication as to whether they had been in a scrap or not. If stunting was indulged in it was safe to assume that they had not been scrapping, but if they landed immediately, it could be expected that they had been fighting. With fuselage struts of not much more than an inch thick in wood, it did not encourage one to throw the machine about if bullets had been flying around. Again the chaps would be eager to get down and obtain confirmation of any enemy shot down and to talk it over. We always knew to within a few minutes when patrols were due back, and always kept an interested look out to gauge the result of the flight.

Sometimes one would hear machines returning well

A fine official photograph of Major McClaughry in front of a Camel. This officer's considerable organising ability speaks for itself in the section he wrote in the Appendix.

before their normal time for return, perhaps singly, and it would be immediately apparent that something had gone wrong. You could sense it, particularly from the manner of their approach and landing.

In Trescowthick's case, he was early and flew in low, made a half circuit of the drome, then came straight in and landed. We were all immediately seized with a cold apprehension; our worst fears were justified.

It was 6th September and the arrangement had been made to rendezvous with a couple of flights of No. 2 Squadron Australian Flying Corps and some Bristols of No. 88 Squadron. These fellows were to assume the above guard position some two or three thousand feet further up. If "A" Flight felt like having a smack at something, the top people would come down and lend a hand if the odds were a bit more than usual. This kind of job had been performed hundreds of times before and had never failed, but this time something went wrong. After some time over the line, Trescowthick saw about three times his own number of enemy scouts some distance away and a little lower down. He checked up and saw that No. 2 Squadron chaps were in position, and so dived into an attack. Unfortunately the people above had a large number of the enemy above them again, and could not come down, but some of these also swung round our "guard" and went in after Trescowthick and his pals. In a couple of seconds they were battling against five or six to one and from a bad position; namely, they had Huns on top, as well as their own level. It did not last long, and four good men were left behind. Taplin was shot about badly, but finished up a prisoner-of-war. Lockley was killed. Eddie managed to land his machine all right as he was seen to walk away from it, but he was later reported killed. It would seem that he was shot later.

My last few days in France were accordingly depressing ones. There was only one thing to do about it so we did it.

Very early on 14th September, I left No. 4 Squadron. From Australia nearly two years earlier until that date, I had been with it almost continuously in its days of infancy when we were all so unversed in flying or fighting, through the period when we started to gain confidence, until we had built up a very fine fighting unit. To walk out of it into what was to be practically peacetime conditions seemed all wrong. I felt less like going to England than doing anything else in the world. The men who had looked after my machine and equipment so faithfully, each one a personal friend, the chaps in whose company I had fought and with whom I had kicked up my heels, were to be left to continue with the war whilst I more or less departed for the ease and comfort of an English countryside. It was with a lump in my throat that I said cheerio to them all that morning. Some of the fellows came with me to Boulogne and saw me aboard for the Old Country. "Blighty" which we all regarded as a sort of Mecca to be visited as often as possible, was now to be my home indefinitely and I was not at all enamored with the prospect. I had had eleven months continuous flying in France and this was rather unusual, but I could have continued.

Human nature is very resilient. A fortnight's leave after reporting at Wing Headquarters, Gloucestershire, and everything soon readjusted itself. England always offered compensations.

Fighting Instructor

CHAPTER 10

I found on taking up the job of Wing Fighting Instructor, that the standard of training had improved vastly. Schools for training and qualifying instructors had been formed, and the much-coveted Gosport "ticket" was something every instructor was anxious to obtain. Gosport was the location of the premier school of this nature, and the highest grading, A.1, was extremely hard to obtain. You had to be much better than normally qualified to get it. So I had to go to Gosport, but first had to learn to fly the old Avro 504K trainer, and learn the ritual of instruction, known colloquially as the Gosport "patter".

We had two stations in Gloucestershire, in addition to Wing Headquarters in Tetbury, one at Leighterton and the other at Minchinhampton, with some ten miles between the two. Many old friends from No. 4 were distributed between these places. Matthews was Wing Examining Officer, a kind of senior staff officer for training. Malley was a flight commander at "Minch", whilst Tab Pflaum and Bill Brasch held similar appointments at Leighterton which was also my headquarters.

One soon drifted into the new manner of living, and the towns of Gloucester, Bath and Bristol saw a good deal of us, and especially Cirencester and its King's Head Hotel with its genial proprietor, Brockman. The local people were all particularly decent to all our fellows. Our aerodrome more or less on the estate of the Duke of Badminton, who was a great sportsman. The Badminton hounds were his private pack I think, but although some of us took part in "the hunt" now and again, we never learned the correct drill. Nor did we try very hard I'm afraid. We were always calling the hounds "dogs" which just isn't done, or getting in front of people who had some pre-emptive right to be ahead of others, and we always regarded the day as a kind of open steeplechase meeting, particularly our Light Horse chaps. Our hosts were always kindly tolerant of our shortcomings, and we were always welcome. They would never let us talk of shooting the poor old fox, however, even when describing such events in Australia.

Matt took me in hand to teach me the "patter" of instruction and also the art, à la Gosport, of flying an Avro, and I found the latter most difficult. Probably the Avro was the most simple type of machine ever to fly, but eighteen months of flying Camels with all their unusual

Cobby as Wing Fighting Instructor. The effect of eleven months of continuous action is mirrored in his face. "The strain of this sort of work was much worse than flying in France."

characteristics, especially with a "war load" on, had so ingrained unusual habits in me, that I found it almost impossible to turn the things properly. It took a couple of hours a day for six weeks to get the hang of it, but eventually I got things straightened out. In the meantime I had selected the best Camel I could find, and had it painted all white, except for a broad black check pattern through it, with a draughtboard radiator cowling and wheels and tail fin similarly adorned. This was not personal conceit, other than the selection of color, as the machine of the Wing Fighting Instructor had to be easily picked out in the air. My job was to "scrap" with all the advanced pupils as frequently as possible and this was a full-time job. Nor was it a safe task by any means. In France one nearly always met opponents who could fly well, and turns were always cleanly made, dives and zooms were controlled almost to inches, but on a training station it was different. You would arrange to meet your pupil at a certain height, and having explained that he had to dive at you, get on your tail, and try and stick there, you would usually discover that the enthusiastic pupil would come too close in his dive. Then if he "sagged" at all as he pulled his machine out, or turned, he had a good chance of knocking your wings off. In the dogfight that would ensue, one had to be particularly careful with the amount of encouragement given to the pupil, because if you allowed him too many liberties, you would find him skidding into you. He would be working on the assumption that you were *par excellence* as a pilot, and would rely on you to get out of the way, but occasionally one's machine would be at its limit in some manouvre or other, and you had to rely on the pupil to do the "ducking". On these occasions only a thread separated you from a collision. On the other hand, if one's full ability was employed to ensure that the pupil in the other machine never got near you, he would imagine he was too far below standard to be able to look after himself and would be discouraged, and the instruction would be a waste of time.

I know the strain of this sort of work was much worse than flying in France, and three or four different mock fights a day, of about twenty minutes or so each, was enough for anyone.

Garnie Malley and I used to collaborate a lot in demonstrating that Camels could be thrown about the air in the closest of flying, without danger, if the pilots knew each other and their machines. We would visit each other and one of us would poke his wingtip in behind the other chap's, then take off. We would then loop, roll, dive, zoom up vertically from the ground level, fly upside down and generally tear the sky to pieces. But we were insistent that it was purely a demonstration and must not be practised by others near the ground. On one memorable occasion, we were stunting round Minchinhampton, I in my black and white checked bus, and Malley in his, which was pure white — when the old man Ossie Watt, approached the aerodrome in his Avro. This also had some gold trimming with the pet name "Papillon" in gold lettering along the side. We immediately tucked our wings in behind his and throttled back and were so close that he would not land. We then looped round him from the side, then from the rear, and then on each side of him just above his top plane. He finally went down and landed, then as he touched the ground we "leap-frogged" over him and taxied in in front of him. When he reached the tarmac he was boiling, but we kept out of his way until sent for.

The usual punishment for low flying, or other such foolish stunts, was overseas for instructors and Malley and I had asked a number of times to be sent back to France. I was privileged in my flying capacity as fighting instructor, but our effort was so blatant, that I was certain I would be sent overseas too. Ossie was in a terrible rage, and was jamming his eyeglass into his eye then dropping it out again as we approached, a sure sign of bad weather to come. We both felt a little jubilant within, although outwardly calm, and then the colonel opened up on us. He could not say "overseas" quickly or emphatically enough, although he tried hard, but our efforts proved a slip. Old Matt, who had been on the tarmac and had seen the show

"I selected the best Camel I could find and had it painted all white, except for a broad black check pattern through it." There were three known basically all-white Camels, the personal mounts of fighting instructors Garnett Malley, 'Tab' Pflaum and Harry Cobby.

from start to finish, interposed and suggested that the colonel should allow him a few words in private. After a while the colonel agreed and a little later we were again sent for. Ten minutes later still, two very crestfallen officers left the office; everything had gone wrong. Matt, the wily old beggar, had told the colonel that in the first place we wanted nothing more than to be sent back to France, and secondly that such an excellent demonstration as ours was of the greatest value in training Camel pilots. A great number of trainee pilots had been killed learning to fly this machine, as its tricks took some learning, although they were really simple to overcome. Its main trouble was that owing to its very small wingspan, and its purposely unstable characteristics, coupled with the gyroscopic effect of a rotating engine and propeller, it flipped into a spin very easily at low speeds. Consequently, in landing and taking off, a tremendous number of fatal accidents occurred, and a general feeling of dislike for the machine was prevalent. It really had people frightened. Matt worked on this and insisted and finally convinced "Ossie" that not only would he be playing into our hands, but our method of flying would help tremendously in convincing everyone that the machines were perfectly safe and were also excellent for fighting. Unfortunately for us the colonel saw the point of the argument and when we went in expecting to be told to pack our bags for other shores, we were smiled at in a most kindly manner and told that we had put up an excellent performance, in fact we were almost good little boys. So that was just exactly that! Then I went to Gosport for the instructors' course for six weeks. I imagined that this was going to be easy, but I was sadly disillusioned. It took something like twenty-four hours to discover how little I knew about the accurate flying of an aeroplane, and the instructors did things with the old Avro that I did not think possible. Major Norman Brearley, later of commerical aviation fame in Western Australia, was my tutor and the manner in which he put that crate down into tiny fields surrounded by tall trees and got it out again, was astonishing. It seemed to be to me impossible yet I thought I could fly. The only way I could have got it in during the first week or so, would be to wheel it through the gate, and as for getting it out, I would have had to adopt similar tactics. The manner in which Brearley sideslipped vertically, right into a corner of the

The cause of this accident is a mystery. None of the aircraft seems damaged but someone obviously is. It is a good descriptive picture and shows some of the facilities of a training squadron.

A fine oblique photograph of Minchinhampton. In the foreground, Aircraft Repair Depot hangars are in the course of construction. Meanwhile Bessoneaux hangars serve the purpose, the nearest five for No. 5 (Training) Squadron A.F.C. and the other five for No. 6 (Training) Squadron A.F.C. No. 5 trained reinforcements for No. 4 Squadron in the field, and No. 6 turned out pilots for No. 2 Squadron. Note Captain Garnett Malley's all-white Camel amongst a flock of Avro 504Ks and Sopwith Scouts. At the other end are Avro 504Ks, Sopwith Scouts and SE5As.

Lower. Leighterton was 13 kilometres (8 miles) away from Minchinhampton and this aerial view shows big permanent hangars under construction and also operations under canvas. No. 8 (Training) Squadron A.F.C. (supporting No. 4 Squadron) is at the lower end of the hangar line. The fighting instructor's Camel is clearly visible. The top half of the line shows RE8 two-seaters of No. 7 (Training) Squadron A.F.C. This unit instructed men as reinforcements for No. 3 Squadron, an artillery and observation unit in the line.

Captain 'Tabby' Pflaum's all-white fighting instructor's Camel with its red band and white emu panels. In the background are RE8s of No. 7 (Training) Squadron, minus Scarff ring rear gun mounts. White aircraft for instructors meant easy identification for pupils in mock combat.

Pflaum's Camel and, from the left, "Mac, Garth, Wilkins, Tab, Mat, Phil". The 'Mat' in this instance is Captain G.C. (Skipper) Matthews, the early "A" Flight commander of No. 4 Squadron.

field, probably facing in towards the trees, and descending sideways at a rate that made it almost impossible to breathe, literally took my breath away. As for picking up how it was done, it didn't happen for quite a while. This crazy descent would continue almost vertically sideways right down to the ground, until it seemed that the wingtip was just about to dig itself in.

Then suddenly the descent would be converted into a smooth turn, level with the grass, and would continue round portion of the perimeter of the field, and then just as speed was almost too slow to stay put we would be in another corner with the maximum diagonal run almost miraculously before us. Then a gentle settling onto the ground followed and a stop well before we reached the far boundary. I heaved a vast sigh of relief. The take-off was another example of the art, only somewhat the reverse procedure, with the treetops brushing the wheels as we crossed the boundary.

After six weeks of sheer hard work, with about three hours a day of concentrated practice under the critical eyes and hands of Brearley, I managed to obtain an A.1 grading, but I have often wondered whether I would have been as successful if I had not just come back from France with a few enemy machines to my credit, and was a wing fighting instructor. Honesty suggests that these facts had a lot to do with it. You could hardly have the No. 1 pilot in the wing on the job with a grading lower than some of the fighting instructors. On the other hand I would not like to think that the way made easy for me. Norman Brearley certainly never let up, and although gentle of voice and undemonstrative in language, he had a way with him of letting you know exactly what he thought of your efforts.

And so back to Leighterton I went, not to give dual instruction, which I appreciated as I did not like it, but to carry on with the job of keeping out of the way of the hard-flying enthusiastic learners of the art of air fighting. Demonstration of rolls, spins, climbing turns, Immelmann turns, rolling off the top of loops, and so on, covered the early lessons. Instruction on the ground was first with chalk and blackboards, then demonstrating it in the air in front of your pupils, then sending off into the air to watch you from your own level, and then you flew alongside them in the air while they tried themselves out, well up in the air. It was interesting only in that it had to be done in order to give them the knowledge to look after themselves as soon as they were ready to go overseas, but after so many months of intense personal activity against enemy pilots, it was very tiring. The lads were so eager for it, too. Any time of the

Upper. Ever since aerodromes have been shared, taxying accidents have occurred. This tangle took place between an Avro 504K (left) and a Sopwith Camel, both A.F.C. training aircraft. Left. Another of the many training squadron Camels lost by accident. The cause is unknown. Lower. Another wreck joins the graveyard of No. 7 and No. 8 (Training) Squadrons. Camels and RE8s await conversion to components or burning.

After a flight, Cobby, in sheepskin thigh boots, walks away from an SE5A of No. 6 (Training) Squadron. The engine is still running and although unchocked, there is probably an airman at each wingtip. One is visible behind the student on the right of the picture.

day or night was open season for questions on how would you do this or that, what would you do in such a set of circumstances, what was the best method of attack, when did you open fire, etc. The questions were endless and, of course, had to be answered. They wanted to know and were entitled to be told, and remembering the complete lack of knowledge with which I started off in France, I gave them everything I could. Particularly important was the art of gunnery. Every cadet was made to develop the habit of swinging his machine on to targets in the air or on the ground, and every opportunity was taken to use ground guns and rifles. Accurate shooting is mainly a matter of training the eye and co-ordinating it with muscles and nerves, and even without a gun or a rifle you can teach yourself a lot.

My two younger brothers, Roy and Jack, were cadets at Leighterton, both having served with the A.I.F., and they were well on their way through their flying training. Unfortunately, although both ultimately qualified and were commissioned, neither got their "wings" as the Armistice came and the granting of the flying badge ceased. It was pleasant to have three of us together; at least the home folk thought so.

Tab Pflaum and I lived in the same block and it was

The SE5A in this photograph is the same as that above. It was unique because it was painted entirely red and was the personal aircraft of No. 6 Squadron's Fighting Instructor Captain Les Holden. Shades of the Red Baron! This was due to the decorative freedom allowed instructors of the 1st (Training) Wing A.F.C. "Corporal Allen of Melbourne", a rigger or engine mechanic, is seen here with the aircraft.

The personal training Avro 504K of Captain Pflaum, illustrated in color on Page 56. Colonel 'Ossie' Watt, O.C. of the 1st (Training) Wing A.F.C., insisted that the name 'Tab' should be on it after suspecting Pflaum of being partly responsible for an earlier low-level raid by instructors on his Headquarters at Tetbury, the halfway point between Leighterton and Minchinhampton.

something like old times, especially that part which included dinner at the "King's Head" in Cirencester. They were most pleasant occasions.

A cause for amusement whilst in the squadron in France had been the arrival on several occasions of a file of papers, asking me to do the impossible in regard to some act of omission or commission on my part in the early days of the war.

Just before the Armistice the same file was again delivered to me. It had been following me since the outbreak of war. It was now about six inches thick and had many hundreds of documents in it. Briefly it related to the mobilisation of my regiment (the 46th Infantry) or the Brighton Rifles, when war was declared. I had been instructed to mount a guard of fifty men at an ammunition factory, and it had not been included in any scheme of rationing, nor did the factory people know anything about us. They kindly gave us a large hall for our use, but we did not have a table or a form, a mattress or a bed, blankets or any provisions. What's more, we had nothing to cook the food in, or to eat it from even if we managed to get the food. So the next morning I had obtained some claim forms from the factory manager and sent a detail into the nearest centre to requisition a hand truck, cooking utensils and the various other items needed for a mess. They also had to obtain meat, vegetables, bread, etc. and I had to guess the scale of rations. Claims were signed for everything for some six weeks, and although I tried many times per day to contact someone at the Barracks who could fix things up, I was unsuccessful. I advised my C.O., who was with the rest of the Regiment at The Heads (the entrance to Port Phillip Bay). I think he was responsible for our relief by a guard from another regiment. So we rejoined the battalion and some months later the file, backed up by scores of claims I had certified, made its first appearance. It was admitted that the procedure I had adopted was quite sound under the circumstances, and that the costs involved were reasonable. All I had to do was to get a quittance by signature, from each men on the guard, for his field allowance of 3/6d per day, the sum of which would be used to help meet accounts. I obtained what signatures I could, but unfortunately a large number of the men had enlisted in various units of the A.I.F. and were in camps all over the country. I duly returned the file with my comments, through my commanding officer.

These papers followed me for 4½ years through various army headquarters in Australia, to the regiment, back to headquarters again, overseas to A.I.F. headquarters in London, through British Army headquarters to France, through various army headquarters to the Royal Flying Corps, then through brigades, wings and ultimately to the squadron. The request was always the same, would I please get the signatures? When it first arrived in France, I was the only member of the list whose location I knew. Many of the others had been in Gallipoli and of course had been killed, but that file of papers actually came to me three times in France and quite a lot of people had given vent to their humor within its hallowed covers. There were poems, sketches, rude suggestions and dozens of quaint remarks within its pages, provoking many an austere comment by someone of senior rank who utterly lacked a sense of humor. Here it was back again, but this time Tab fixed it once and for all. After putting in a week reading some of the brighter portions, he said he would sent it on for me, and I never saw it again. When I asked what he had done with it, he said he had stitched it up in aeroplane linen, bound it many times with cord and sealed every crossing of the cord with sealing wax and addressed it on a metal plate to the "King of the Hedjaz", with that Monarch's full title and address. He had then put it in the official mailbag for Australian Headquarters in London and I have never had sight or sound of it since, nor has anyone raised the question of its whereabouts. Perhaps the many claim forms have been used as a sort of "shin-plaster" currency in Irak or Palestine for the past twenty-two years, but whatever happened, it has not come back. I am still haunted, however, by a vague fear that some day my children, or my children's children, will be asked by the generation of claims' experts of the time, to obtain the quittances by signature of some fifty members of the Brighton Rifles in the year 1914 A.D.

Early in November came whisperings of a possible Armistice. The station wireless office immediately went onto a continuous watch and everybody studied the expression on the faces of newcomers to a room or the Mess. For days every ring of the telephone, no matter when it occurred, brought an instant cessation of work and noise. Everyone would go tense and after a few moments, if no excited yell burst from whoever answered the telephone, we would all relax and get on with the job. We knew that there was a general retreat of the enemy in France, that there had been some form of mutiny in the German Navy and that their army was fed up, so we were expecting something to break. Continuous rumors were being passed around and everyone was on edge. Then it came. A crowd of us had been up all night in the wireless office, as we had received information that German plenipotentiaries had crossed the line, and when the news came through we made one concerted rush for the mess. But supplies of the stuff that cheers were very scarce, and to gain the proper effect quickly, many weird and potent drinks were mixed. Then it was off to the "King's Head" and from thence to London.

London during the Armistice! The first week nobody went to bed. Hotels, restaurants, all were open house. The streets were a seething mass of dancing people all day and night. Old gentlemen, factory girls, ladies of high estate, soldiers, sailors and airmen doing impromptu Sir Roger de Coverleys everywhere. The lid was off and one suddenly realized that everyone had been living under a great strain. It was all over.

The Armistice left us all somewhat up in the air. It was not necessarily the conclusion of the war, but a period which, if everything worked out correctly, would lead to peace, as indeed it did later. We knew there were few ships available to get the troops home again. We also knew that until all the conditions of the Armistice had been satisfied, the armies would have to stand to their arms, but with nothing to do, how would the time be spent? We felt that the period would be a difficult one, as idle men — especially large numbers of them — become impatient and resent any form of restraint. Luckily for us, limited flying was permitted, and this kept everyone busy to some extent, but a gradual easing up of control became noticeable. Parties of cadets and officers would ask for leave for periods that a few weeks before would never have been contemplated and it was granted because there was little reason for staying on the aerodrome. People would vanish by aeroplane for days at a time, and on their return would recount their adventures around England. But continual leave has its shortcomings, and unless one is blessed with many friends, to return home without something in the pocket until next pay day, is hard. We were, however, fortunate in the type of man in the A.F.C., and although people often became edgy, nothing of the slightest consequence occurred. A good deal of drill was indulged in, although hitherto, other than falling in and marching to and from the hangars, it had been almost unheard of. It was accepted wth a grim sort of cheerfulness as one method of filling in time.

For me the time passed pleasantly, but I had to be doing something. London obviously saw a good deal of me, until finally I went into hospital to have some very troublesome tonsils removed. Wandsworth Hospital was very handy to my London "home" at my aunt's, and during a convalescence which covered a very elastic period, I saw everything I could of the old country. I stayed on the farm

Upper. Headgear as worn by the well-dressed training instructor, modelled by Cobby's old chum, 'Tab'. Centre. The 'Richthofen Circus' aircraft were brightly painted but some of the 1st (Training) Wing aircraft were also colorful and at times hilarious in their paint jobs. In this case Pflaum's joey kangaroo was firmly attached to the 504K's petrol cap. Lower. Tab and a pupil leave 'Joey' astride the Avro's fuselage.

Another of the many accidents. A neat little Sopwith Scout of No. 8 Squadron on its nose in the middle of Leighterton aerodrome.

Big brother to the Camel, a Sopwith Snipe is seen here trestled for examination at Leighterton. Preparations were under way for Snipes to replace the Camels in wing fighter flights but the Armistice made this step unnecessary.

of friends of my people and even took a couple of weeks at Bournemouth and surprised everyone by swimming several times a day. This was a happy period, but it had to end. Soon I received a week's notice to report back to Leighterton. We were due to sail from Southampton as a shipload of air force personnel in a few weeks' time. It was now mid-April 1919 and everyone had been given pre-embarkation leave prior to sailing.

We had been informed in a roundabout way that there would be a march of the A.I.F. through London on 25th April to commemorate the landing of Australians at Gallipoli in 1915 and that we could march if we so desired. But we were on leave and did not so desire. On that morning the Australian Flying Corps was probably scattered all over the British Isles, but a large proportion of them were still in London. About ten in the morning, word was passed around that the A.F.C. would be allowed to fly over London during the March on this, the first Anzac Day. The Prince of Wales was going to take the salute accompanied by General Birdwood, in front of Australia House in the Strand. The "word" also said where you could obtain the type of machine you wished to fly, near to London on one of the several aerodromes in the vicinity. So until lunch time the clubs and the pubs were full of convivial pilots of the Australian Flying Corps on leave, with the war over, and on a date in the near future, we were to sail for home. A perfect setting for a swan song, but an event that was to remain in the memories of many thousands for a long time. At about twelve o'clock, Australian pilots began to take the last one or two for the road, and then make their way to their destinations to pick up the machines they proposed to fly. London was a sitting shot, everybody was merry, there were no restrictions as far as we knew and it promised to be a gala day for Australia. We had won the war and could now bid farewell to Blighty, where we had had such a glorious time. One could not think of anything more fitting nor anything more stupid than the authority that permitted it.

The head of the march would reach Australia House at 2.30 p.m. At 2 o'clock odd aircraft started to appear over the Strand, getting lower and lower. At 2.30 about fifty or sixty machines were jockeying around the sky over Australia House, and more kept on arriving. Aircraft were looping and rolling and spinning everywhere, but the libations of the morning had added more verve than sense to the manoeuvres. As the head of the column approached the saluting base, there was a mad scramble to pass the Prince, and some of us went down into the Strand in order to do the thing properly. It was probably the most foolish thing I have ever done, but having got down, I had to stay

"On 6th May we embarked at Southampton in the S.S. *Kaiser-i-Hind.*" This photograph shows the vessel at Port Said.

down for about three quarters of a mile before I could get out again.

Somebody else in a Pup, well known to those who participated, also got down, but fortunately some distance behind me. If he had been in front, we would have been involved in a frightful smash as I was in a Camel, which was of course much faster than a Pup and I would have run him down in a couple of hundred yards. The overhead wires from building to building kept us down and I would not have been able to avoid him, and with the roads packed like sardines with onlookers, the casualties would have been terrible. Finally I was able to zoom up just short of Trafalgar Square, but by that time all I wanted to do was to get back onto the ground as soon as possible, and I did. That first Anzac Day will live long in the minds of many people. The other chap got out too — but how I don't know.

On my return I found advice from wing headquarters that I was to proceed to Minchinhampton for the last few days as they were short of senior officers, so I packed my bags and made off to my new station.

On 6th May we embarked at Southampton in the S.S. *Kaisar-i-Hind* for Australia. It was a damp but striking send-off. It rained continuously all day and the many distinguished visitors who came to farewell us, including Sir William Birdwood, were entitled to added decorations for turning up. In particular, the mayor and corporation in their robes and chains of office plus and the city band, put up quite a good show. They stayed on a dais, making speeches and waving hands for a couple of hours, while the troops on board tried to throw oranges down the bell-mouthed ends of the larger brass band instruments.

As we moved away from the wharf the crowd sang Australian songs and we on board joined in, and although we were all cheered at the thought that it was all over and we were going home, there were mostly sad faces on board. We had had such a wonderful time and been so hospitably treated that there was no pleasure at all in that moment of leaving England. Sadly, every one of us had left staunch and close friends behind.

Officers, wives, and ladies on their part of the vessel, and in the background, other ranks.

These men all returned home. Upper. Captain Les Holden (Nos. 2 and 6 Squadrons). Upper right. From left, Captain Garnett Malley (Nos. 4 and 6 Squadrons), Captain Edgar Davies (No. 2 Squadron), Major A. Murray Jones (O.C., No. 2 Squadron), Lieutenant Norm Trescowthick (No. 4 Squadron) and Captain G.C. Wilson (No. 2 Squadron). Right. A game of deck tennis in progress. Lower. Fittingly, this is a photograph of the Australian Flying Corps leading ace, Captain A.H. Cobby, enjoying the trip, on this occasion in the company of Sister Tyson, one of the many nurses also returning to Australia.

The inclusion of this final section in addition to Cobby's autobiography provides important insights into his narrative. It is an extract from "Australian Airmen", a history of No. 4 Squadron Australian Flying Corps, compiled by Lieutenant E.J. Richards. This section was written by Major W.A. McClaughry, seen here in the cockpit of his Camel.

Appendix

Squadron Organisation in the Field.

(By Major W.A. McClaughry, D.S.O., M.C., D.F.C.)

An Aerial Scouting Squadron is made up of the following chief sections:—

Three Flights—Each having eight machines.
Squadron Workshops. Motor Transport.
Technical Stores. Armoury.
Orderly Room Staff.

Each Flight is commanded by a Captain (Flight Commander), who is responsible to the Squadron Commander for the condition and serviceability of his machines, and the trainng of Pilots in his Flight. The Flight consists of about thirty-two men, including—One Flight Sergeant (Chief Mechanic), Two Sergeant Mechanics (one fitter and one rigger), and Four Corporals (two fitters and two riggers). These N.C.O.'s are so distributed that a fitter and rigger is entirely in charge and responsible for each machine, the N.C.O.'s supervising generally, and each having to overlook a certain number of machines.

In No. 4 A.F.C. I arranged a system whereby all technical work was done in a central squadron workship. This included all engine repairs, and work for tinsmiths, coppersmiths, blacksmiths, oxy-acetylene welders, car-penters, sail-makers, and electricians. To enable me to get the best results from this scheme, I withdrew from the Flights a number of the best tradesmen, including a Sergeant and Corporal, who worked under the orders of the Squadron Technical Warrant Officer.

Thus, any work required to be done—apart from actual adjustment of engines in machines, and rigging of machines—was carried out in the Squadron workshops. All engines which required overhauling were removed from machines, and the work was done in the workshops, a new engine being supplied from the nine spare engines carried by the Squadron, which were kept in the workshops. Ordinary top overhauls of engines were carried out by Flights, assistance being rendered by Squadron workships if required.

All new machines that came to the Squadron required many additional fittings and gadgets to meet the requirements of the various pilots. Included in such fittings were new gun controls, gun-pads, extensions on Aldis sights, additional wind-screens, bomb racks, etc. These would be made and fitted by Squadron workshops. The workshops were also responsible for the upkeep and running of workshop lorries, which included the plant for electrically lighting the camp.

The Squadron establishment of motor transport was seven Leyland heavy lorries, two Leyland lorries, five Crossley light tenders, seven flat trailers, one Crossley touring car, five motor cycles, and two side-cars. The transport was supervised by the Squadron Equipment Officer, a sergeant being in charge of the vehicles and responsible for their upkeep and regular overhaul. Squadron Standing Orders laid down that one transport vehicle must be laid up each week for complete overhaul. No. 4 Squadron was unfortunately, on proceeding to France, issued with very old transport, which was withdrawn from other squadrons. Great difficulty was always experienced in obtaining spare parts, tools, etc.

The transport allotted to the Squadron was totally inadequate to cope with the work of transferring the Squadron to a new aerodrome during a move, but a Reserve Lorry Park was attached to each Aircraft Park, where sufficient spare lorries and trailers were kept in reserve to be loaned to squadrons detailed to move. If only one squadron happened to be moving at the time, plenty of spare lorries were available, but when several squadrons in the same Wing were moving together, the allotment of spare vehicles was very limited, and sometimes totally inadequate.

The care of all machine guns is a very important branch of work, and in a Service Squadron sufficient emphasis cannot be laid on the necessity for a keen and efficient Armament Officer. He is entirely responsible for the condition of all guns and gears and also the training of armourers and gear mechanics. There were four armourers on the strength of each Flight of 4th A.F.C., working under an Armourer Sergeant, and these attended to the Vickers guns with which our machines were fitted. A Gear Sergeant was responsible for the condition of all interrupter gears, which fired the guns through the propellers. To assist the sergeant it was found necessary to train two gear mechanics in each Flight.

The Squadron was originally equipped with Sopwith Camels, fitted with Kauper gears. These are extremely delicate, and require constant adjustment and attention. Extraordinary good results were obtained from these gears throughout, and I attribute this almost entirely to the efficient work of the armourers and gear-fitters.

Afterwards we were re-equipped with Sopwith Snipe, fitted with "C.C." gears. After getting over the initial troubles arising through lack of experience, these gears also gave entire satisfaction.

The testing of ammunition, and filling of belts, was a matter which required very great care, as each individual round had to be tested before filling into a belt. Special mechanics trained to the work were engaged in this. I made it a practice that all new pilots coming to the Squadron should receive instruction in guns and gears under the personal supervision of the Armament Officer, and certain tests were laid down which must be passed before a pilot could be given leave. These tests included a general knowledge of the mechanism and working of the Vickers' gun, and the type of interrupter gear in use at the time. Also clearance of stoppages on the ground and in the air.

The work-benches, cupboards, etc., used in the Armoury were all made in such a manner that they could be fitted in a heavy lorry, for use on a move, should the Squadron be required to move forward quickly and work without proper accommodation. I would like to make it clear that I attribute much of the success of the Squadron in no small measure to the very efficient manner in which the Armament Officer carried out his duties.

The Squadron Equipment Officer is responsible for the indenting, obtaining, and storing of all technical stores; also flying clothing and equipment. The practice in France was for all squadrons to keep an absolute minimum of technical equipment on hand, in order that rapid packing and movement could be assured. An Aircraft Park was attached to each Wing, where a plentiful supply of all necessary supplies were carried, and these could be obtained immediately required by wiring through Wing Headquarters. Two Leyland lorries, two flat trailers, and one wing trailer were permanently allotted to the Equipment Officer for use in case of a move, and all stores were fitted in special bins fitted in the lorries.

Including the Flight Commander, there were eight pilots in a flight, each pilot having his own particular machine. In order to make pilots take the greatest interest in their machines, they were allowed to have distinctive markings or colourings painted on them, or to affix special mascots.

Flight Commanders must train their own pilots, correcting all faults in flying, such as fast landings, flat turning, etc., and prevent foolish and dangerous stunting near the ground. They must take the deepest personal interest in each individual pilot, so that they thoroughly understand the temperament of each, and they must insist on all Standing Orders and rules being read, understood, and carefully carried out. A keen inter-flight rivalry should be fostered. The efficiency of a Flight Commander can be quickly and easily gauged by the behaviour of his pilots; by their flying, or the conditions of their machines; and chiefly by the standard of his formations in the air. I always insisted upon good formation flying; chiefly because straggling formations were more liable to be attacked unawares than a good one, and inexperienced pilots lost.

It is imperative that a Flight Commander should be keen, and absolutely full-out to crash enemy machines, or attack ground targets from low altitudes, as he is entirely responsible for the leading of his Flight, and pilots will quickly catch his spirit. The success, or otherwise, of a Flight Commander can only be judged by results.

Flight formations were always led by the Flight Commander, or his deputy leader, who should be an experienced pilot, selected because of his dash, keenness, and ability to lead. Whole squadron formations would be led by the Squadron Commander or senior Flight Commander. These consisted of the three Flights intact, and led by its own leader, following one commander. Usually on normal occasions it was found that a Flight of about six machines was ample for an ordinary offensive patrol or escort work; but on occasions of great hostile activity it was often found necessary to patrol in squadron formations, comprised of upwards of eighteen machines, or even Wing formations of four or five squadrons. On the latter occasions, each squadron was intact as a formation and flew together at slightly different heights, following a selected leader, who might be the leader of any one of the squadrons.

I always made a practice of personally interviewing new pilots at length, and going fully into all details of the work they would be required to do, and strongly impressing upon them any

A squadron on the move.

difficulties likely to be met with. New pilots would then be posted to a Flight, and before being allowed to participate in a patrol over enemy territory were practised for upwards of a fortnight in fighting in the air, formation flying, and firing at targets from the air. They would then be taken by an experienced pilot to view the line and learn the country; and before being included in a patrol, would have to find the way around the whole of our sector of the battle front alone. Meanwhile on the ground, their training consisted of passing the gunnery tests before mentioned, obtaining a thorough knowledge of the contents of the Intelligence Hut, and a series of lectures on all subjects of interest connected with their work, i.e., enemy tactics, formation flying in the air, etc.

The Operation Orders, detailing patrols to be performed by the Squadron, were issued from Wing Headquarters each evening. These orders would then be issued to Flight Commanders in typed form, giving them times of patrols, number of machines to be sent, whether bombs were to be carried, patrol area, height to be flown, etc. Should the order include anything unusual — such as whole squadron formations — a conference of Flight Commanders would be held to work out a scheme. Or if an escort were required for artillery or bombing machines of another squadron, the Flight Commander concerned would be instructed to personally interview (or arrange over the telephone with) pilots of the artillery or bombing machines, regarding exactly what was required. Such details as time, place, height of rendezvous, and various Verey light signals to be used, must be carefully considered and fixed.

During periods of normal activity I always considered it safe for experienced pilots to work singly or in pairs in the early morning or late evening for the purpose of making surprise attacks on enemy artillery machines and balloons, or ground targets, such as trains, hangars, billets, and troops and transport on roads. In this manner extraordinary good results were obtained with practically no loss to ourselves.

In a service squadron but little time can be spared for such details as drill. But my usual practice was that on every morning parade at 8 a.m. about ten minutes would be devoted to smart drill. On days when flying was impossible, longer parades were held, and I also found it necessary to order special classes of instruction for N.C.O.'s. A syllabus was arranged, the course being one hour daily, extending over a period of one month. Sergeants would attend one month, and corporals the next.

A guard of one corporal and six men would mount over the camp each evening at sunset. A fire picquet, of one sergeant and twenty men, was detailed weekly, and none of the members of this picquet were allowed to leave the camp while their tour of duty lasted.

Separate messes were provided for sergeants and air mechanics. All forms of outdoor and indoor games were provided by a sports fund, and whenever accommodation permitted it, a special hut was set aside for these indoor games. A regimental canteen was also established, where all possible necessities and luxuries were sold at the lowest possible rates. The profits accruing from this canteen were used partly to provide sporting material and games, and partly to provide extras in the air mechanics' mess, such as vegetables, oatmeal, flour, fruit, etc.

In order to be successful it is absolutely essential that a spirit of keenness and cheeriness should pervade the Squadron, and grumbling and cantankerous officers should be quickly removed. This cheerful spirit should be fostered in the air, at work, in games and sports, on the ground, and in the mess. The officers' mess should always be comfortably furnished, well lit and cheery, a piano and gramophone being absolutely essential. The whole atmosphere of the mess should be so arranged that officers cannot help taking a pride and interest in it and in the squadron. Good meals should be provided, even if it is necessary to hire a cook from outside the squadron. Whilst all against heavy drinking in the mess (a failing which is all too common), I do believe that when the occasion arises to celebrate some special success of the squadron — such as a big bag of Huns, or a decoration for a pilot — this should always be celebrated in a proper fashion, and a special dinner arranged with speeches, toasts and music. This was our invariable custom in No. 4 Squadron. To my mind, the spirit of a squadron is the chief factor upon which success and good results depend. Once traditions have been established, the work of maintaining the required spirit is much simplified.

Enemy Aircraft Destroyed

Pilot	Destroyed	Out of Control
Capt. A.H. Cobby, D.S.O., D.F.C.*	24	5
Capt. E.J. McClaughry, D.S.O., D.F.C.	21	2
Capt. R. King, D.F.C.	16½	6
Capt. H.G. Watson, D.F.C.	8½	5
Capt. T.C.R. Baker D.F.C., M.M.	8	4
Lieut. L.E. Taplin, D.F.C.	8	4
Capt. G. Jones, D.F.C.	6	1
Lieut. A.J. Palliser	6	1
Capt. G.F. Malley, M.C.	5	1
Lieut. T.H. Barkell, D.F.C.	4	3
Lieut. N.C. Trescowthick	4	1
Lieut. G.S. Jones-Evans, D.F.C.	4	—
Lieut. E.R. Jeffree	4	—
Lieut. A.E. Robertson, M.C.	2	2
Lieut. F.J. Scott, M.C.	2	2
Lieut. P.J. Sims	3	1
Capt. J.W. Wright	1	3
Major W.A. McClaughry, D.S.O., M.C, D.F.C.	2	1
Capt. E.F. Pflaum	1	2
Lieut. H. W. Ross	1	2
Lieut. R.G. Smallwood	3	—
Capt. W.B. Tunbridge	1	2
Lieut. A.W. Adams	1	2
Capt. D.P. Flockhart	1	1
Lieut. A.E. James	1	1
Lieut. A.H. Lockley	2	—
Lieut. W.S. Martin	1	1
Lieut. G. Nowland	1	1
Capt. A.H. O'Hara-Wood	1	1
Capt. N.L. Petchler	—	2
Lieut. A.D. Pate	1	1
Lieut. O.B. Ramsay	2	—
Lieut. F.S. Woolhouse	1	1
Lieut. H.A. Wilkinson	1	1
Lieut. R.H. Youdale	2	—
Lieut. V.H. Thornton	½	1
Lieut. H.H. Browning	—	1
Lieut. J.C. Courtney	—	1
Lieut. E.A. Cato	—	1
Lieut. T.R. Edols	1	—
Lieut. A.T. Heller	—	1
Lieut. O. Lamplough	1	—
Lieut. R.F. McRae	—	1
Lieut. R. Moore	1	—
Lieut. R.C. Nelson	1	—
Lieut. A. Rintoul	1	—
Lieut. E.J. Richards	—	1
Lieut. V.G.M. Sheppard	—	1
Major F.I. Tanner	1	—
Lieut. J.H. Weingarth	—	1
Lieut. J. Wilkinsons, M.C.	—	1
Lieut. L. Wharton, M.C.	—	1
Lieut. H.N. Kerr	½	—
"A" Flight	2	—
"B" Flight	2	—

*In addition to this tally, Cobby shot down 13 German field observation balloons.

INDEX OF AIRCRAFT ILLUSTRATIONS